Dead Green Roots
Greens and Failed Socialism

by
K B Napier

Petra Press, PO Box 415, Swansea, SA5 8YH   UK

Dead Green Roots
Greens and Failed Socialism
by
K B Napier

ISBN: 978-0-244-13031-2

Petra Press, PO Box 415, Swansea, SA5 8YH   UK

Foreword

The foundation of the Green movement is Marxist socialism.

Marxism is the foundation of Fascism.

The foundation of the global warming movement is Marxism,
wielding a fascist baseball bat. Now, 'Greens' have morphed 'Global Freezing' into 'Global Warming' into 'Climate Change'. But, the aims are the same!

The ones wielding the bat are Al Gore, rampant socialists, and opportunistic governments who covet your cash and obedience.

Greenism is a destructive political unscientific wing, very different from genuine environmental science. It wants to destroy capitalism & Christianity, and to gain worldwide power. It has nothing to do with the environment.

This book has been taken from the first three chapters of my longer book, 'The Global Green Agenda: Second Edition' (Lulu, 2009), sold by Amazon, Lulu, and bookshops. Its intention is to prove that the Green movement is just a devious attempt to revive the old failed socialism of Russia, Nazi Germany, and other ruined countries. Though first published in 2009, the principles and facts have not changed, so the warning implicit in its pages remain also. Don't be duped!

To put it bluntly, only the rich have benefits from the Green movement. Everyone else pays through their collective nose while their living standards decline. On top of that a major aim of Greenism is to cut the population of the world by any means possible, whether by imposing crippling taxes on the poor who use 'fossil' fuel, by unchecked diseases, or by stealth. It is also about making huge gains for the 'leaders' of this nasty movement. As in the old USSR, they lead a great life, while those who actually do the work and pay for their rich lifestyle suffer.

Listening to Greens is like purposely drinking gallons of cancer cells. In 2009 I proved that the claims made by Greens were lies and deception. Nothing has changed. Greenism is bad science conducted to maintain big departmental funding for unscrupulous scientists. Now, they are trying to stir-up more trouble and the same old call goes out – "We will all die within ten years unless…".

This is EXACTLY the same call they made ten years ago (read the 2009 book) when they called it 'global warming'! Ten years before that they called it 'global freezing'!! Real scientists know all this. Everyone is being hog-tied so they can't fight back. They'll take your money and allow the Third World to die. (It's all in the 2009 book). If you are happy with that, well, there's no point in talking about it.

I decided to publish just the first three chapters because Al Gore and his super-rich mates want people to die, and are making another attempt to deceive, this time swapping 'global warming' for 'climate change'. Frantic Greens yelling out that we are all going to die unless we pay their ransom money is ludicrous! THINK HARD!! If you allow them to continue, even more radicals will take over. This is their desire. Is it yours? If not, fight back and drive them back into their dank caves.

Preamble

Important Distinctions

I first started to write about this topic in global hotel publications, where critics and pro-Greens quickly attempted to smear me with the usual epithets.

But, what was I saying? The same as I am saying today, which is:

Greens and pro-Greens (including hoteliers) do not know the difference between good management and environmentalism. So, everything they do that should simply be called 'good management' is invariably dubbed 'Green'!

However, their inappropriate use of words is not just down to ignorance (which it is). It is also down to cynical exploitation of people's emotions, who are equally ignorant of the facts. The facts are these:

There is a need to use resources wisely and properly. This should be a given, but, sadly, many who are in a position of management rarely manage. That is why they now rely on 'Green' this and 'Green' that. Not to save the earth, but to make more profit and capture the sad folks who are 'Green' as customers.

Good management is the appropriate use of resources, reducing or eliminating waste, trying not to use chemicals unless they are absolutely necessary, not to buy new things unless the old ones have become useless or shabby, and not to use materials to extinction. Staff should be sufficient to do the job, things that don't need mending should not be fixed. Marketing should not exploit the fears and emotions of people – though many advertising campaigns do just that. All these things are simply management issues. They are NOT 'Green'! I am 100% supportive of genuine management, because it is good stewardship of finite finances and other resources for the reasonable financial upkeep of the business, and for the reasonable provision of goods or services to customers. Both should match, with no exploitation by either.

But, I am NOT in favour of environmentalism, or 'saving the planet' (I cannot think of a more twee and worthless marketing strap-line!). This is because everything 'Green' is about political power, and not about good management. Recent events in finances and politics prove this to be true.

Good management is about optimum use of a limited number and type of resources. 'Green' is about manipulation of people to bring about a political end, regardless of financial or other considerations, in order to gain maximum power.

'Green' involves the use of futuristic models and 'what if' scenarios. Good management looks at concrete facts and figures before making decisions. 'Green' mixes all kinds of esoteric influences, from paganism to untested (or failed) economics. For Greens everything is about the future... 'IF this happens then we will have to do this, or that.' And it is all smothered in socialism.

Worse than that, Green movements want to spread their ideas by dictatorial propaganda and legal constraints, forcing everyone to obey their mantra. They cover their aims with mysticism, Mother Earth nonsense, and unsustainable arguments about sustainability. Green, then, is unnecessary and political, and is *not* the good stewardship of sound management practices.

Greens use mythical 'problems' about resources as excuses to manipulate others. They use language designed to be vague and without substance. Behind it all is a hatred for capitalism (the very system that gives them the money to attack capitalism and which provides necessary money for keeping people alive) and the Christian ethos. The outcome of following a Green agenda is a downward slide into financial, political and social instability and eventual destruction.

There is, then, no rational reason to support or listen-to this false environmentalism. It is a Marxist theory coupled to neo-fascism. That means, as with the two older failed historical versions of the same movements, that a few want to manipulate and control the majority. If you are 'Green' then you are already press-ganged into obedience and service, just like peasants and farm-workers before you.

Environmentalism, as I will prove in this book (Note; This refers to the 2009 book, 'The Global Green Agenda'), only exists because the media and governments are refusing to give two sides to the story. Only the President of the Czech Republic spoke the truth on this matter, later to be joined by President Trump (who is far too conservative and level-headed to be allowed to continue, according to Greens!).

The summary, then, is that we should encourage attempts to act with a sense of good stewardship, but NOT out of obedience to 'Green' philosophy, which is unbalanced, extreme, deceptive and dictatorial. Because Green equals failed socialism (Marxism and fascism).

Green has nothing to do with 'saving the planet'; it is everything to do with creating an elite Marxist-fascist society.

I urge you all NOT to use the terms of this political movement: 'Green', 'environmental', 'save the planet', 'CO2 harm', 'climate change', 'global warming', etc. Once you use these terms, you spread the disease of 'Green' further and the truth is hidden even more.

Say 'Yes' to good management of everything.
Say 'No' to anything Green, foolish and unscientific.

Contents

Chapter 1

Marxism
and Environmentalism

His hand hides the smoking gun
that will be used to bring the world
to its knees long after his death

To many hopeful but naïve folks, the fall of the Berlin Wall, November 10th 1989, was the beginning of the end for communism. But, it wasn't. A concrete wall is not an ideology, and ideologies do not fall as quickly or as easily as a wall. And, communism has many faces, using different disguises and words. Even in the West, whose peoples fought hard against Marxist deception and atrocities, this is not apparent. Yet, the USA is now (2018) in civil meltdown as Democrats (Marxists) try to violently overthrow the legitimate President and government. If they succeed, the fools who vote for them and support their wicked thugs, will soon come to realise how stupid they have been, because their freedoms will be demolished and they will be forced to live a lesser life under communism and fascism combined.

Soviet communism might have staggered a little in 1989, but its children fight on, sometimes in the most unlikely of movements... Terroristic Islam, homosexuality, UK government, the European Union, the UN, US Democrats, local governments, environmentalism and more, including a variety of 'Christian' groups. They all exist because the first wave of violence did not seem to be violent – Political Correctness, the staple and backbone of all socialist iron-fisted rulers. The fake president, Obama, made sure he began a violent reaction to freedoms, because he is both Islamic and Marxist/fascist (Islam itself being fascist).

It is important to recognise the communistic source (shared with fascism) of the present 'green' movement, so that it can be opposed with some confidence.

The old Soviet Union regime dissolved in August 1991. Then, in December 1991, Russia, Ukraine and Belarus, joined together as the Commonwealth of Independent States, moving the capital from Moscow to Minsk. At about the same time, Dr Bo Hi Pak addressed the 20th American Leadership Conference in Washington DC to give a speech outlining the "deep, spiritual reasons behind the failure of communism." (1)

Dr Pak, referring to watching the live TV report of the falling Berlin Wall, said "We doubted our eyes and ears". Yes, we did. And so we should have, for the Wall was only a symbol of East versus West, not the underlying tyranny versus freedom; communism versus the free world (and free market). At that time I knew communism would not die, because those with murderous and totalitarian intent never give up. Sadly, I have been proved right. Worse, socialism is gaining ground by violence and victim societies just roll over and accept it because of 'diversity'.

Marxism allows for rapid acquisition of power and wealth for the few without going through the usual freedom processes of voting or acceptance of ideology. That is why people who cannot muster support by any other means, will simply drive a truck over anyone who opposes them (literally, in the case of Islam). It has been said very recently by academics in US universities: 'Why bother to argue your case when simple force will do!' Indeed,

Clinton openly said this in 2018. Thus, the PC movement has replaced truth and the tedious bother of explaining its position, with blunt power without genuine authority; and those who reject what has been declared will be removed, or made to appear uneducated, or simply an 'enemy'. Or, in Green parlance: 'Deniers'.

The fall of the Wall and the old Soviet regime was indeed swift. But, like Nazi leaders who quickly donned the clothing of ordinary soldiers and peasants when they were beaten by the Allies, so communists took on another persona. Those who loved communism did so because they gained something by its presence, whether it was money, food, status, power, or all four. Plus the opportunity to take revenge or just hit out in hatred at those they did not like – Christians especially. Everyone else gained nothing and could only laugh and cry when the yoke was removed. Die-hard communists had to find a new face behind which to hide their oppressive ideals.

As you read what Dr Pak said, look at the fake environmentalism now spreading its green tentacles over the world, and recognise it as a child of Marx. The methods used are the same, as is the aim. Dr Pak said: "A wave of democratisation has swept over Eastern Europe and even the Soviet Union itself." Now, in 2018 this has been replaced by the fascism of Islam and the EU.

No, what swept over the Soviets was a wave of acceptance that the old way was now gone. The wave was not deep enough to purge the hearts and minds of communists who wanted to retain the power they once had over others (which is why modern Russia now sees an upsurge in communistic support). Nor did the wave wash away the desire to control and be oppressive. This is because violent oppression is a result of man's sinful nature; it is the lowest common denominator.

> "Let us ask why communism has failed. For 70 years communism has been a great menace to human freedom and to the very survival of our way of life... In the past 70 years since the Soviet Union established a totalitarian state, their claim to legitimacy has centred around the belief that Marxism-Leninism was the scientific truth. They had absolute confidence that science would prove them right – that only communism was scientific and that science would relegate God and religion to mythology. During this period, the one word that bolstered communism again and again, was the word 'scientific'... In fact, the most fundamental reason why communism failed is that it was neither scientific nor the truth."

Dr Pak was right in his description but was wrong about communism dying. It did not die, because some men love to lord it over others, and know that to do so effectively, at least for a season, requires a revolutionary imprisonment of their minds and bodies. This must be done quickly, brutally and without remorse, with no room for argument or dissident voices. This is exactly how Greens and socialistic governments are now acting! If anything their nasty resurgence is becoming greater and more violent. Just look at the USA right now!

In addition, you must control how people live, what they say, and what they do (except for yourself, of course). And this is what we see today in fake environmentalism and other totalitarian movements, including homosexuality and terroristic Islamism. Marxism is the language leaving Al Gore's lips, and naturally permeates the Green movement. Pak continues:

> "The communist philosophy stands on the supposedly scientific assumption that the essence of the universe is matter. Only matter exists and is real. The eternal spirit, or soul, is said to be an illusion. This is the foundation of dialectical materialism, the metaphysics of Marxism. In this view, there is no room for God. To Marxists, God is really an entity that man created, not the other way around."

Intriguingly, even non-Christians bring God into any discussion of Marxism and fascism, as if there is a residual realisation that something vital is missing from every large political and social movement. The Soviets hated any notion of God, and so they terrorised and tried to eliminate Christians and their faith. They continue in this evil reasoning in China and other socialist countries.

This path of elimination is still found in China, Mexico, Africa and other places. It is also arising in Europe and the USA, where all trace of Christianity is being eradicated systematically and ruthlessly, where it is okay to be pagan or atheist, or even terroristic Islamic, but not Christian.

Exactly the same thing is happening through the dictatorial policies written for the world by Al Gore. It is not my imagination, because it is what actually Gore says. The point is plain; it does not matter if you are a Christian, or any other brand of believer – Al Gore does not want you or your ideas. He just wants Marxist dictatorship.

Christians, and any other believers in God (or 'god'), are blamed by Gore for the supposed demise of his aged mother, Earth, so they must be eliminated or at least silenced. Marxism and fascism are fine vehicles for doing that.

It is very interesting to note that Alexander Solzhenitsyn received a Nobel Prize in Literature in

1970 for his Gulag books, depicting actual horrors of Marxism, and yet, in 2007, Al Gore was given (I will not stoop to say 'earned') a Nobel Prize for Peace.

It is interesting because anyone who has read every detail of Solzhenitsyn's books, not just the extensive Gulag's, will not recognise anything of peace in Gore's Green movement. The 'peace' part rests on the validity of his (now debunked) theory that we are doomed to environmental disaster that will bring worldwide war if not dealt with. So, by dealing with the 'problem', we avert war. Hence, 'peace'. It is like the farcical 'peace' promised by fascist Islam – but what it means by 'peace' is a time when everyone is subjugated by Allah! That is not peace. It is totalitarianism.

That the logic behind this does not exist, and there is no 'problem' to begin with, is of no consequence to the Green movement, so long as they impose their way of life on everyone. And it is this enforcement of ideas that really rattles my intellectual, moral and spiritual cage!

The Soviet existence was filled with doom and gloom, disaster, financial death, personal imprisonment for being dissident, and 'diagnoses' of criminal insanity for believing in God, or objecting to socialism. This is now on the rise again in Europe and the USA.

There were two levels of people: the very few who led (by force) and those (the majority) who were commanded to follow whether or not they agreed to it. The leaders had control and did what they pleased, being unhampered by their own rules or civilised activity; 'the led' had to put up with it and suffer. Even believers in communism suffered. Look at Solzhenitsyn's books in depth (because they are written that way) and you will not only see something from the past, but what is promised to us in the future by Al Gore and the movement he spawned, which was kept alive by Obama.

The bleakness he says is coming is not caused by anything done by man, nor by scientific mishaps, but by lack of both – and men stepping aside and letting Gore and friends rule the earth. Al Gore is the disaster – not any mythical earth problem! And again in 2018 he sees his goal again, because the majority do not oppose him.

It is with great force that Gore demands we act now, without thinking. This is because thinking will generate suspicion and questions about the truth of his claims. (And slow down the wealth he and his risk-capital friends hope to gain rapidly. See later notes in the 2009 book). These potential and present dangers to his proposed regime must be silenced very quickly or the West will be doomed by socialism.

Well, he certainly LOOKS like Obama

Look at him. Look at his movement, now adopted by Obama and the Clintons. See what it really is: Marxism rising from the ashes. See the parallels between modern environmentalism and past Soviet rule, because fake environmentalism/Greenism is Marxism resurrected. And, as we shall see in the next chapter, it is Marxist tyranny imposed by fascist techniques.

There is no need to laboriously search for these facts, because Gore and the Green movement are very open about the need to smash anyone who stands in their way. By law or by social outcasting, they will crush dissident voices. The Kyoto conference used fairly normal but impassioned language. But, the later Bali conference had an edge to its language: brash, arrogant and fascist, taking no prisoners. The signs are there – start worrying!

Solzhenitsyn spoke of the danger we in the West are facing, by not acting against communism. (2) He said (and he should know!):

> "Until I came to the West myself and spent two years looking around, I could never have imagined to what an extreme degree the West had actually become a world without a will, a world gradually petrifying in the face of the danger confronting it... All of us are standing on the brink of a great historical cataclysm, a flood that swallows up civilization and changes whole epochs."

Not long after that speech was made, the UK had a Marxist socialist government which, under open Marxists in the shadow government, is still fighting to gain power, and the USA fell to the Democrats, who are Marxist through and through. So, the sudden presidency of a Republican (Trump) came as a massive shock to Democrats (communists) who tried every dirty trick in the book to gain power. Most in the USA thought Tony Blair was a great Prime Minister. But, many UK citizens think differently, seeing him as a puppet of the EU, and a destroyer of all that was British.

Part of his policy was to allow illegal immigration to run rampant, for one reason only – to destroy British culture and way of life, making it easier for the EU to take sovereignty, which it did, and which the UK is now trying to destroy. Tony Blair, then, was not a good PM; he was, and is, a very good Marxist, who bludgeoned the people into silence and servitude.

If anyone should understand the universal evil represented by communism, it was Solzhenitsyn. And, his prophecy turns out to be true and fearsomely imminent. For Solzhenitsyn, the disaster was rooted in agnosticism and atheism: "the calamity of an autonomous, irreligious, humanistic consciousness." Even for those who prefer a world without God, this holds true, because these empty beliefs engender violence and freedom to be bad.

To put it bluntly, a country ruled with genuine Christian laws will show compassion, help for the poor and unwell, training in what is good and true, freedom of expression and 'small' government. A country ruled by godless people will be just that – godless, in all aspects of its life.

I am not just talking about people who do not believe in God. I am talking about those who make their unbelief an excuse to deliberately bring in laws that allow them to do whatever they like, usually damaging the majority in the process. In 2018 we already see a complete rule by homosexuals and a growing rule by Islamists (both are socialistic in ideology).

These are nasty individuals who do not wish to discuss their views or actions. And they will flout their 'scientific' beliefs even when such beliefs have been derailed and intellectually dismantled.

To be godless (whether spiritually or politically) is to invite disaster: morality, health, economies, ideas... all will suffer. They discovered this in the Soviet Union, and the same strands of devastation are found today in the three major movements already mentioned – homosexual, Green and extreme Islamic. They all want to obliterate any form of godliness and force their regimes on everyone, though what they offer (no – demand) is demoralising and filled with injustice and horrors.

I keep repeating it: anyone can believe anything they like, but they may not force me to believe the same things or do what they do, and may not harm others just to satisfy their own whims.

This is a prime directive of communism. Once communist rule has been achieved, a country slides fast into a pit of its own making, without hope or light of day. We see this today. I urge you to read the Gulags for proof. Or, look at current communist regimes. Do you really want to live under them?

Solzhenitsyn ran to the West to get away from communist darkness, and found the West was itself running, but towards the very communism he was constrained to run away from! It is one thing to be a zealous, vocal, Young Socialist in the backroom of a Western pub or bar, but a very different thing when you actually live under Marxism.

As with boys who dream of being gangsters, it is somehow romantic. But, when they join an adult gang and experience the fear, death and hatred, it chokes their souls. Right now, Gore is romancing the dream of communism, as are those who follow him. (The same idea of 'romance' was also found in Nazi Germany).

Soon, if he gets his way, he will be a leader amongst men, with all the freedom he has designed for himself, but which he denies to others. His followers are sycophants who are given rewards for perpetuating his demands. Everyone else will be oppressed and crushed. Anyone who rises up will be put away and silenced. It is not fantasy; this process has already started and the whole policy is in proposed Green laws.

As Solzhenitsyn said, communism:

> "made man the measure of all things on earth – imperfect man, who is never free of pride, self-interest, envy, vanity, and dozens of other defects. We are now paying for the mistakes which were not properly appraised at the beginning of the journey..." (3)

The same social revolution that gave birth to the failed Soviet Union is the very same revolution calling everyone to accept the myth of global warming/climate change.

"Capitalism's inherently anti-ecological nature results from its subordination of the needs of society to the accumulation of profits, regardless of the costs to society as a whole." (4)

This is the talk of people living in a 'capitalist' country, and it is very simplistic! Those who actually live under communist rule dream of the day they, too, will join a capitalist country! It is absurd to say that all who own businesses do not care about anything but profit. Even if they did, it is a fault of persons, not of the system.

Profit is needed to keep business alive. It pays for wages and everything we (including western dreamer-Marxists who have never lived under oppressive Marxist regimes) take for granted: health provision, pensions, benefits, housing, good food, education, freedom of speech and movement, and so on. It is hardly, then, making people's needs subordinate to profits. It is profits that fuel the many advances in society.

When people work for themselves they produce profit and workers are paid a wage. When people work for a system that allows for no personal gain from hard work, they just do what is necessary and, as they found in the Soviet Union, workers could not care less, and only do what is minimal*. This is because, in effect, they are just slaves without rights. In that Soviet atmosphere, prison, illicit gain, black-marketing, poor finances and planning, ill-health, and massive fiddling of the national books, were all rolled into one. Rooted in the same rotten principles, this is what fake environmentalism means for the world. (Note: There is a genuine environmental science, but it is not found in the Green movement. In this book 'environmentalists' refer to the fake kind).

(*The Founding Fathers of the USA discovered this for themselves. At first they worked almost like slaves for native Indians, and their demise was rapid. But, when they took control of their lives and began to grow their own crops and work for themselves, they had an incentive, and flourished. This is where the U.S. work ethic came from, together with its freedoms, so easily overturned nowadays, by the Obamas, Clintons, Blairs, UN and EU of this world).

Marx's theories enjoined environmentalism. So did Hitler's Nazi Party. Both forces are naturally a part of environmentalism. It is about time the world was shown its true environmentalist parents!

One weapon used by aggressive modern environmentalists is the simple one used by all dictators who love Marxism: propaganda, using lies. This is being opposed by real environmentalists themselves. For example, Bjorn Lomberg (who Al Gore refused to talk with. Gore also refused to give him scientific reasons for his stance. That should say it all, especially as Gore's views have been proved to be lies and political posturing).

Political, But Not In Politics

"The Green Movement is puzzling people today, particularly when it takes the form of a Green Party, and most particularly in connection with the German party, by far the most important one, Die Grünen. They are said to be unpredictable and unable/unwilling to make any compromises with any other actors on the party-political scene; consequently they are not really in politics, they are only political." (5)

Galtung makes two assumptions about greens: "The Green movement is an umbrella movement for a number of partial movements, each one of them attacking one or more elements..." and "The Green movement differs from other social movements in denying that basic social problems can be solved attacking one single factor; a much more holistic approach is needed."

These are correct assumptions, and this is why we see green movements today wanting total control over whole economies (Or, rather, dis-economies!).

Green policies can be summed up as: 'going back to nature'; regulating society through small pockets of control (which also refer back to a centralised body); cooperative businesses, usually without profit; complete vegetarianism or near; alternative technologies; self-sufficiency; the insanity of de-coupling from super-powers (the basis for being overrun), believing their simplistic theories will protect them from attack (Obama messed about with this idea); production with a barter system; non-formal education; feminism & sexual 'freedoms' (together with their social and health destructions); elimination of dirty or manual work (so, who's going to do it?); living closer to nature, though distant tribes would dearly love to get away from it; less predictable lifestyle; enhancing sex, health foods, animal rights, etc.; own entertainment. "There is a correlation in the ideological universe".

You will note the claimed decentralisation of power and authority, even though reality makes us look at societies throughout the millennia, where, despite desires to be independent, others of a less altruistic nature invade Utopia and occupy it!

In this 1980's observation, the author sees Greens accepting the existence of Christians, whereas today there is a hatred for all things tagged 'God'.

. This occurs because of the addition of 'Gaia' to the mainstream ideas of Greens. Thus, Marxism itself is changing in form, but only in peripherals.

Strangely, Galtung seems unaware of the Marxist roots of environmentalism, for he says "The Green movement, with its focus on non-violence, is also a rejection of terrorism and single-factor Marxist determinism." Even though everything environmental has a clear stamp under its belly: 'Made by Marx' and 2018 Greens will use any violence in their power to gain control.

He envisions a Green movement that does not seek power as such, but empowerment in matters they wish to change for themselves, or for others of like mind. (Animal rights and feminism have arisen from Marxism, and are far from non-violent!).

This is not what we are now seeing, as Greens have evolved, their wickedness getting the better of themselves and their claimed 'ideology'. Rather than be empowered within their own kind, they wish to impose draconian regimes and laws on all who resist. They have quickly replaced their flowers of love with shackles of dictatorship. They do not care if everyone follows because they accept the ideology; straightforward obedience will suffice. This is exactly the view of Islamists – they do not care if people actually worship Allah, so long as they submit to Islamic rule! This is blatant fascism (or Marxism - choose your poison; both are socialism).

Ignoring the different formations within Green parties, we can see Marxism throughout. In a book examining the difference in structure between European and USA/Australian Greens, we therefore find four sub-headings, the last being "the different contextual roles played by the anti-nuclear movement and wilderness experience, and ecology, Marxism and the new left." Marxism is there, not just in black and white, but in the very heart of greens. (7)

Today's Greens Still the Same

> "The regulatory role of the Confucian scholar in feudal China was to warn the Emperor and other officials of the landlord ruling class, when their squeeze on the peasantry became so severe that revolt might be provoked. The best of these scholars were quite sincere in their belief in Confucian ethics. This, however, did not negate the fact that in social reality they were acting as a stabilising mechanism to maintain the system of landlord oppression."

Curiously, much of the Green Movement can be understood in the same way...

"While the Confucianists bird-dogged the landlord class for its own long-term good, the Green Movement can be seen as bird-dogging the modern industrial capitalist class so that its need for profit will not undercut its necessary foundation in nature. Thus, the warnings on global warming, resource depletion, biodevastation, and so on... the regulatory function of the Green Movement is to warn of this danger." (8)

The article this quote was taken from makes allusion to Marx, as it would by necessity.

This kind of talk reads as benign and welcoming, but the reality is far from benign, being based on enforcement, hatred for people (see later notes), and control of people's lives by unnecessary taxation, curbing of energy, denial of progress to poor countries, legal penalties, social castigation of dissidents and loss of freedom of speech, coupled with intellectual deception using bad science. These will all be proved as the (2009) book progresses.

Greens Mix Reality with Fantasy
Back in 2001, one author correctly identified a big problem with greens: their ability to mix real problems and solutions with myths and fantasy, usually by sprinkling their statements with pseudo-science. Six or seven years later, this became

mainstream activity, as we saw in Al Gore's infamous public meetings, where he continued to teach bad science that has been debunked. But, it does not matter so long as the Green imposition remains on track. In 2018 he again uses the same tactics.

"It's not unfair or inaccurate to say the Green movement has most, if not all, of the elements of a religion, and only a tenuous relationship with science. Of course, uttering such heresy will outrage Green warriors.

Mother Nature has again, as in animist times, taken on the aspect of a goddess. She is seen as normally gentle, but capable of the vengefulness of Yahweh should mankind in general, and scientists in particular, dare try fooling with her. As Gaia, she's ready to intervene in everyday conduct. While they have no proof these things have much real effect on the visible world, they do find these actions spiritually rewarding, and paying lip service to ecology does fulfil a human need.

Traditional religions place Jehovah, God, or Allah first. Greenism places the earth first." (9)

Since the article was written things have moved on, very fast. Gaia worshippers now want the total destruction of anything involving a personal God, including Christianity and all who believe in a personal God/Jehovah/and even Allah.

They have a pathological hatred for these religions because they believe a personal God is devoid of interest in the planet and, as Al Gore insists, their religion is "responsible for the spoiling of Mother Earth". The answer is therefore to get rid of the religions and those who believe in them. For a while, the Green movement appeared to be benign. But, in just 20 years, it became a people-and-religion hater.

The same article continues:

> "The Green movement has many structural similarities to Marxism... It assures adherents that history is on their side... Both Marxism and Greenism have diagnosed the world's problems and offer solutions that are not only psychologically appealing to many people, but also seem morally 'right'. Like Marxism, Greenism pretends to be science, but amounts to dogma."

> "Now that Marxism has been relegated to the scrap-heap of history, its adherents have to find a new centrepiece for their

belief system without losing the Marxism. And just as the Marxists had a hidden agenda of controlling other people, it is arguable that most of the professional Greens do also. After all, at least in the West, they are largely the very same people. Protecting the earth… serves as an excellent pretext for almost any controls. This opens the door, as with the Marxists, to an 'end justifies the means' approach." (See later chapter on Transition Towns).

"It has to be said that the movement's roots are planted in ignorance as much as in malevolence. The same was true of many socialist sympathisers… Especially towards the end, socialist beliefs had only a limited intellectual appeal, since the facts contradicted them everywhere. Their appeal was psychological and spiritual. In other words, religious and not susceptible to reason."

"Most Greens are non-violent, but the movement is increasingly strident, and there is also a strain of malevolence and suppressed violence… long ago the equivalent of a Leninist wing arose… and ran around spiking trees and monkey-wrenching bulldozers. The Greenpeace

'Green Warrior' performs a similar function at sea."

"The Marxist agenda succeeded in destroying the economy of half the world and a lot of the environment as a bonus. The Green agenda promises to do the same for the environment, with the destruction of the economy as a perverse bonus."

"You may be wondering whether I am insensitive to the fate of the earth and its creatures. Far from it. Protecting the earth as a pleasant place to live is critical. And clearly there are real problems that need to be dealt with. But the solutions proposed by nearly all the Greens would aggravate the problems. Their books almost uniformly advocate political activism and socialist planning... There are much more effective ways to address the problems: more economically effective and more scientifically effective. But not with Green science or economics."

As I say, things have moved on very fast since 2001. But, the author above got it right at the time. Green is Marxism under another name. It is amazing that the USA, so hard against Marxism in the past, is now embracing the very same perverse politics that have ruined so many countries, persisting in building its own 'Hanoi Hilton' to punish dissidents with. All that will happen is that the rich get very rich and the silenced-poor will get poorer. However, they will have nice (worn) dungarees and lovely potatoes, but no electricity and oil, and no choices.

Confusing Propaganda and Public Concern

Robyn Eckersley began researching his book on greens and politics in 1985. At that time, he says, Green political thought was

"relatively new, relatively underdeveloped, and (a) reasonably manageable field of inquiry. By 1991 the literature on Green political ideas had expanded rapidly – a development that reflects the increasing international public concern over environmental issues..." (10)

The author does not understand that the huge increase in Green ideas is not a 'reflection' of "widespread public concern", but a reflection of how successful Green propaganda has become! If Greens had not spread propaganda, very few people would have this 'concern' over things they do not really understand, and have never thought about! Nor would they jump up and down crazily over pseudo-science. Sadly, they have become 'willing pawns' in a communist game.

As is said in the introduction to this book, environmentalism, Islam and homosexuality are all a new expression of an emergent revamped Marxism. This is not my guess, or my political prejudice; it is easily identified in the literature, words and actions, of those involved.

The three groups use exactly the same terms and the same bullying techniques. They are deliberate Marxism, not unintended copies. And Marxism pushes the limits continually, knowing that somewhere, sometime, the limit will break and it will gush forward like a flood of destruction. We are now almost at that dangerous point.

The worldwide 'acceptance' we see now for Green ideas is not acceptance of the facts at all, but acceptance of the propaganda given out by Greens, because of the immense pressure and governmental laws forcing them to 'accept'. That makes it a very different thing.

To suggest that people everywhere have a clear knowledge of what Green ideas really mean, is to say all those people subdued by Stalin were willing slaves. In Marxism you are told only what the leaders want you to know. If you try to find out more, you are a troublemaker and will be incarcerated or silenced.

In mid-2009, Barack Obama stopped people finding out whether or not he was eligible to be President of the USA. He refused to provide a birth certificate, and it appears that he entered the USA as a student from a foreign country, and was listed as a foreigner. His own grandmother says she was present at his birth in Kenya!

Naturally, we all wanted to know if he had any right to the Presidency... but, he put armed security forces where files may or may not be found! He even sent them to Kenya to stop people asking questions. This is because he is not only a liar, but a Marxist... shut up or suffer the consequences! He applied the same iron-fisted rule to everything, including Green policies. By 2018 everyone now knows Obama WAS a fake, and still is, and his birth certificate was fraudulent.

Like so many, Eckersley is naïve. This is evidenced by his use of the phrase "The environmental crisis..." to begin chapter one. He has resorted to the classic 'If – Then' argumentation model, in which the second proposition is reliant on the proved status of the first proposition, even though there is no proof for it in the first place. This is very bad procedure and allows for big lapses of logic and truthful data, and a negation of genuine science.

He does, however, discover that Green thought can be divided into several separate but connected forms, "under the broad generic names of eco-Marxism, which includes both orthodox and humanist eco-Marxism... Ecosocialism, and Ecoanarchism." (11)

Eckersley concludes that no one Green ideology constitutes a cohesive political theory. And so Eckersley continues with "... the intractable nature of environmental problems..." We cannot be 'intractable' about something that is not a 'problem'! Environmental 'problems' only exist in the prejudiced, conniving minds of environmentalists, not in reality.

Going back to when 'Green' started to formulate a new approach to politics, Eckersley, in common with so many people today (including most 'Greens'), confuses items such as toxic dumps, nuclear plants, pesticides and pollution, with the current movement that brings a whole new dimension to their thinking – Gaia. Saving people from pollution, etc., and rethinking how to be industrial without being self-harming, is admirable.

But, to then use all this to 'save the planet' is plain insanity. There is no logic behind 'save the planet', and it moves concern away from helping people, to becoming servants of the cultic earth 'Mother', even though it means the death of millions. In this sudden switch, people are hated by Greens because they are seen to be pollutants to be rid of. (Of course, Greens do not count themselves as one of the pollutants!). It is a stark fact that the inmates have taken over the asylum.

The only ones to 'recognise' so-called 'environmental problems' are those who invented them in the first place – and I seriously question their ability to think straight. Are they insane, as some critics say? What else do you call educated people who worship dirt and rocks as 'mother', and who are willing to kill off most of the population of the planet to 'protect her'? ('Wicked' is a word that comes to mind).

In other times these groups would be diagnosed mentally ill, or at least cranks. Now, they are making a complete turnaround by calling everyone who does not accept their bizarre asylum ideas, 'deniers'. The book alleges that "serious stresses... are clearly visible ahead" and "people will be poorer in many ways than they are today".

The author wrote these words before 2000, so he could not foresee that the 'serious stresses' are caused by Greens, not by any mythical environmental 'problem' that is bad enough to ruin the whole earth! He certainly did not see those unsightly 'wind farms' that have ravaged the UK skyline and sea, and which have almost no impact on energy supply.

Nor could he foresee that people will become poorer, not because of these 'problems', but because Greens are imposing their will on countries, stopping them becoming wealthier by using fossil fuels, etc., and by bringing in legal barriers to stop their improvement.

And, even worse, Greens now want to demonise babies and dramatically reduce world population (that is mass abortions). Not for any good reason. Just to satisfy whims and ideologies that prove themselves to be utterly devoid of truth and real solutions. Marxism died. What we are seeing now is a new form of it, but it is still a dead body, kept alive by a global warming of its flesh by blind, dead men who are strapped to their saddle, lifeless, like El Cid. (In my 2018 book, 'How the World was Made', I refer to the 'dead donkey' carried on the backs of evolutionists).

Marx's Idea – Dead But Warmed-Up

> "… socialised man, the associated producers (must) govern the human metabolism with nature in a rational way, bringing it under their collective control, instead of being dominated by it as a blind power; accomplishing it with the least expenditure of energy and in conditions most worthy and appropriate for the human nature." (12)

Apart from the fact that some producers are greedy, Man has done just this for millennia. Then, along came Marxists and almost destroyed the ability of men to live (as well as the will to live), let alone become better off (except for the leaders, of course).

Now, the newly reborn Marxist movement has moved far beyond rational thought, and is subduing all people to its own ideology, used as a cover for protecting mother earth. Because, just like their former communist colleagues in Soviet Russia, *et al*, these Marxists want personal control over people, not the planet. They want the power it gives them, along with the privileges given to leaders who make up their own rules to live by.

> "The experiences of the previously communist societies are hardly a ringing endorsement for bringing human society's interactions with nature 'under collective control'. The ecological and social problems associated with the Soviet and East European regimes were, if anything, even worse than those experienced under capitalism." (13)

The author thinks Marxism is a good starting point because it can combine social justice with environmental justice. But, what is really happening is that people are sidelined and even hated, whilst Marxists use the environment to gain power for themselves. 'Environmental justice' is a nebulous and fairy-like concept, because it makes 'nature' the new humanity! And 'social justice' in Marxism means totalitarianism and oppression of the people.

(The rationale behind this 'justice' for everything from next door's pet mouse to the 'worth' of cannibalism, is relativism, where everyone and everything is equal. It is a mindless and intellectually inept idea, but it is used by people who want to get their own way for their own cause and purposes).

> "Modern environmental thought... can go well beyond the most obvious critiques and connections between Marxism and environmentalism... Marxism can provide profound insights into the ways in which societies relate to the environment."

Yes, and this same Marxism has already failed in spectacular fashion around the world! It cannot say it did not have enough time – it had over 70 years in Soviet Russia and is still wreaking havoc in nations around the world. Cuba is just stagnant, with no future in its present Marxist condition.

And this Marxism is the new hope being foisted on us by environmentalists today! Can you not see the dangers? The very real decay caused by, and within, Marxism? The route backwards to bad times and the destabilisation of nations?

The current idea is to apply Marx's ideas on 'Metabolic Rift' to today's environmentalism. (14). That is fine – but dead is dead! Marx is dead. So is his theory. Why can't people just live as nice people, instead of as clones of debunked theorists, out to do-down everyone else?

Well, Was Marx a Green Marxist?

Though modern Marxists use Marxian philosophy to repress and oppress everyone who does not accept the environmentalist argument, Marx himself did not actually use his ideas to promote green ideas. "For the early Marx the only nature relevant to the understanding of history is human nature... Marx wisely left nature (other than human nature) alone." (15) This is "regarded by most socialists today as laughable", implying that Marx was involved in environmentalist arguments.

Modern ecologists feel Marx was too tied to industrialisation ('Prometheanism') to be a true environmentalist, and so did not "leave a significant ecological legacy that carried forward into later socialist thought, or that had any relation to the subsequent development of ecology." (16)

Some argue that Marx could not have had anything relevant to say about ecology because he was born in a different era, that knew nothing of nuclear fission, modern chemicals and the very word 'ecology' (17). This is untrue; German thinkers were into 'ecology' before Marx started to write.

Yet others claim Marx had the fundamentals in place and knew more about the relationship between man and nature than ecologists do today. (18)

Foster then comes down to the real point, one that escapes many who observe Al Gore and his movement. It is that technology itself is not the issue, but the nature and logic used in capitalism as a mode of production, is. "Socialists have contributed in fundamental ways at all stages in the development of the modern ecological critique." Marxism is the foundation of modern environmentalist thought!

So, the argument is that Marxism is fundamental to the "devastating environmental conditions that face us today" (though not proved to exist by science), and only Marxism is capable of developing a new world order based on tackling capitalism.

In other words, environmentalists have no genuine interest in the environment at all; their true interest is in removing capitalism. If you do not believe me, just read all the literature put out by environmentalists. They want to remove YOUR way of life, with its relative comforts and income. In 2018 this was made abundantly clear by 'Greens', whose main aim is the destruction of capitalism. The environment has nothing to do with it!

The ultimate thus far in this line of reasoning is the 'Transition Town' concept now racing to gain prominence in the UK and USA. It wants us all to return to a feudal system of bartering, taking ourselves away from the international scene and its links to capitalism, and to become small communities that are self-reliant. In reality, even in small communities there are men who want more, and want status or power.

If people wish to follow this route, that is fine. But, communism has never worked. In the USA the Pilgrim Fathers tried it, and almost died, many of them servants to local Indians. But, once they held land of their own, they flourished. (19)

In this utopian and impossible dream, no regard is paid to the fact that men are prone to be greedy and violent. Some will quickly assert their dominance over the rest. And, others outside these communities would be quick to detect foolish passivity, invading them and making them subject to their demands. That's what history proves.

Yet, the modern author hearkens back to the "Leibig-Marx connection". Leibig asserted in the mid 1840's that big farms were "part of a larger British imperial policy of robbing the soil resources (including bones) of other countries." They did this by growing products in one place and shipping them to another. Today, the same general argument is used to deny Third World farmers the opportunity to sell their fresh products to the UK today, but the detailed argument is that the shipping itself is a problem, because it generates CO_2.

The fact that these poor farmers will die of starvation does not bother socialistic Marxists. They are more concerned with their particular slant on economies. It is hard to support such an idea. In an ideal world men would each grow their own crops and obtain whatever else they need by being thoroughly decent fellows.

Such fellows are rare, especially today. There are others around who dream of great power, and inflict immense hurt or death upon those they want to subjugate. Stalin did this, using Marxism. So have many others since, from homosexuals to Islamists, liberals and the 'left'. The result for thoroughly decent fellows is filth, corruption, war, starvation, disease, oppression, intellectual decline and death.

Leibig argued that Britain grew strong health-wise by 'robbing' other countries of their earth nutrients. Perhaps Leibig should check the health of Edwardian working classes. They were not as healthy as he supposed. And how could he tell, without proper records? No, his was a conceptual cause, not an actual one.

Today, Marxists and others are basing their arguments on concepts and not on actual cases. That is why there is no science behind the claims made by environmentalists. Unless it is corrupt and fraudulent.

The same man decried big farmers and large-scale farming, but it is this very thing that feeds huge populations! It is the science behind large-scale farming that helps Third World people to survive. But, Marxists want to dismantle it all, so that farming and other food production fits their theory against capitalism. That people die is of no consequence. Indeed, their deaths are encouraged. (For proof see the 2009 book, 'The Global Green Agenda', Second Edition, from which this book is taken).

Karl Marx, living in London, was influenced by Leibig's book when he was in the middle of writing his 'Capital'. He was happy to enjoy living in the country he chose to call names! (2018: The same duplicity is used by Muslims who have inundated the West – they hate what it stands for yet take everything it can give). Marx wrote to Engels to say that Leibig's ideas were "more important for this matter than all the economists put together." This is why Marx alluded to Leibig in volume one of 'Capital'. (20)

One of the arguments used by Leibig, for example, was that Britain imported many tons of bird droppings from Peru and bones from Europe, to use on land as fertiliser. "In essence, rural areas and whole nations were exporting the fertility of their land." This is false logic.

The reason locals exported their bird droppings or bones was that they earned good money from it; more than they earned by using it themselves on their own land. It was not 'robbery', but astute business sense by British farmers and foreign locals. From this false episode came the concept of 'sustainability'.

Do you think the Peruvians would export these items if they then starved because they could not fertilise their own lands? Hardly. This is how Marxists and environmentalists construct their arguments; on their own feelings and theories, not on real-time situations and a more dynamic thought process. This was the free market in operation! And Marxists hate it, because their utopia is one of centralised control within a huge government, oppression, and the making of slaves to feed their system. Everyone else can starve and die!

In the given example, Peruvians needed an income; British farmers needed (natural) fertiliser; so they came together and made a contract. The Peruvians did not lose out, because they had money plus bird droppings of their own.

And so along came the idea of 'the law of restitution', which meant that any minerals, etc., removed from the land had to be replaced by the same things. Of course, the obvious question must be: why remove it in the first place if you have to replace it? Hence the modern demand: do not remove it in the first place. Leave it there, even if it can be used to good effect elsewhere but is left unused where it is. Marxism really is that stupid!

This is behind the idea of leaving abundant oil and coal reserves* in the ground, rather than removing it for use, to maintain jobs and to spread wealth. The theory comes before the lives and livelihoods of people. (*These reserves are enough to last the entire world many centuries... and this is based only on the reserves discovered thus far. Every year new fields of oil and coal are discovered, but socialists deny them being utilised, just to satisfy their perverse ideas).

It is Marx who said that Capitalism created an irreparable rift in the metabolic interaction between human beings and the earth, and what is taken out must be put back. According to his theory, growth of capitalism and large-scale food production, coupled to sending food long distances, makes the rift even worse.

That this creates waste is evidenced by the waste and pollution in towns after food is eaten. Marx noted that the waste was pumped into the River Thames, making it a veritable pollution-waterway.

That was true at the time, but we now deal with waste properly and the river is nowadays clean. Marx saw this as 'antagonism between town and country under capitalism'. Therefore, farmers had to go back to small-time production to feed only those within their locality. (21) Marxists, then, are living in the past, and 'see' capitalism and all it represents in the past tense. They refuse to face the facts of modern societies in which new and increasing technological advances ruin the basic tenets of their creaking faith in Marx.

Perhaps I am being too picky, but do you see flaws in his argument? Firstly, despite the continuance of some who prefer personal greed to general good, capitalism has produced the food needed by their countrymen. Yes, it makes a profit, but that is because few of us today can grow our own food. Indeed, the socialists who run governments prevent us from doing so, through taking of land, imposition of foolish health and safety laws, and growing, crippling taxation, and so on. Land replenishment is now fairly commonplace.

So, what is Marxism trying to do? It is trying to send us back to systems that died away, of necessity, before the industrial revolution took place. Watson, of the Sierra Club, wants us all to live in newly constructed communes of less than 20,000, each commune separated by wasteland! The Transition Town movement (22) in the UK (and also in the USA) is very similar in outlook. Transition Towns are so favoured, because they fit the future plans of the vile UN, and the European Union (EU), themselves people-haters and draconian Marxian-fascist monoliths.

The European 'Union' is a collection of similar perverse minds, without the support of the people; real 'support' comes from individuals who are given the true facts and conclude decisions based on those true facts; the EU does not give truth or facts, but mainly it acts secretly and imposes its will on everyone. Therefore it is fascist.

The minds of its leaders are those of powerful men who make inordinate, abusive incomes by ruling through the EU, and subjugating people of many nations to its will. Tony Blair, for example, rejected any idea of coming out of the EU, even though this was the will of the people. But, as a Marxian-socialist, the will of centralised government came first, though socialism damages every country in its path.

By forcing member nations to accept its rule, the EU is dividing-up each nation into smaller regions. Smaller communities will be exactly what the Marxist EU needs... easily controlled small units that have no real connection to each other with peoples so mixed as to lose any idea of national identity. That this policy (multiculturalism) destroys local communities and nations is not relevant to dictators.

It is more efficient to control 20,000 in one village than millions in a whole country. The old vision of 'divide and conquer'.

We do not need Marxism; we need a sound national will to help ourselves to be better, and politicians who are truthful and genuinely acting on their voters' behalf (No, do not laugh in derision – they must be forced to do so!). This must be by personal acceptance, not national imposition of will by a particular political creed.

Have not modern Marxists realised why their earlier political parents were cast out of power? It was because people had enough of being controlled and oppressed.

Marx and Engels "did not stop with the soil nutrient cycle". They also tackled deforestation, desertification, climate change, the loss of species, pollution, industrial wastes, toxic contamination.... Yes, Marxism is the head of false environmentalism.

Marx had a theory to propose. It took iron-men with ice in their veins to carry out his ideas: Lenin, Stalin... people who were willing to murder thousands upon thousands until everyone submitted (or else).

Al Gore genially points out facts and figures on his board to large audiences. The facts and figures are fake, without scientific substance. And his geniality is superficial. Listen to his tone. Watch his face. His inner hatred for mankind is starting to show through. All he wants is his personal fortune and power!

Marx, too, thought he had 'science' on his side. So did Stalin and Hitler. Geniality soon turned to violence and oppression. You have seen it in the movies and in books. Some of you fought against it. And many died because of it. Sadly, as that older generation dies away, so newer generations forget, so the rotten cycle of Marxist-fascist totalitarianism again starts to rule. One day its evils will be renounced – but not until it has run its course and destroyed whole nations.

Why, then, are so many of you swallowing Marxist environmentalist garbage? It is this system of lies and deception that is a pollutant, not capitalism.

Marx and the Good Old Days

"... we need to recognise that Marx and Engels, along with other early socialist thinkers, like Proudhon and Morris, had the advantage of living in a time when the transition from feudalism to capitalism was still taking place..." (23)

Marx looked back on the 'good old days' from the stance of his comfortable life, courtesy of capitalism. He then looked at what was arising and did not like it. So, he wanted the transition to reverse life and culture, even though increasing wealth began to improve everyone's lives, despite the greed of some.

Yes, capitalism was still in a raw state, but after its initial indulgence in greed, whole nations emerged better off. Sadly, even with these facts amongst us now, modern Marxists want to take us back to feudalism, with all its reversals of benefit to mankind.

And, interestingly, the 'good old days' they wanted to revert back to were only 'good old days' in the minds of romantics who lived later, and those who owned land and had inherited wealth! Peasants and labourers who lived under feudalism hardly had a love for their condition! Marxists are not realists!

The 'irreparable rift in the metabolic interaction' between people and the earth is just a fantasy that suited Marx's ideas. The advances in science and agriculture, though imperfect, are continuing to find solutions to any 'rift'. In the process they are feeding millions and enhancing health beyond any possible solutions Marx may have proposed. Yet, his rancid followers still want to impose a failed system on everyone in the world, whilst they retain whatever they want for themselves.

What Marxists are now proposing through the environmental movement can be likened to a man undergoing an operation. He is given an injection to put him to sleep and told to count backwards. Very soon, he is unconscious, knowing nothing, doing nothing.

That is how Marxists want us all to obey, as they take us backwards through time to an era when men fought their own demise, poor health, lack of education, poverty and starvation, dying at an early age. But, they make it all look pretty and romantic, hiding their desire to be our leaders, or, rather, our despotic dictators.

Marx said that the agricultural advances we see today were 'impossible' because of capitalism. (24) But, it is the very capitalism he despised that now creates vast prairies of wheat, and massive fields of all kinds of food. Small independent farmers no longer survive because the demand for food is so great. And science works hard to enhance farming techniques. Marx knew none of this, so many of his theories are null and void. We have moved on!

But, Marxists have not. They are still deluding themselves into thinking that the "metabolic relation between human beings and the earth, (points) beyond capitalist society to socialism and communism." (25)

Communism was tried in Russia: it began with genocide, ruthless violence and hatred, and social collapse, and continued that way until the country was close to its own demise. The same happened, and still happens, in China and elsewhere.

It does not matter where communism exists, it is always kept alive by oppression, denial of freedom of speech and ideas, and death. Communism is a failure and an indictment on all who practise it.

The socialism so admired, has ruined modern Britain, and continues to do so, as Christian laws are replaced by laws favouring the few and the corrupt and the perverted. The socialist government in place in 2008/9 was Marxist, denying freedom of thought and voice, crushing religious dissent and complaints, allowing the presence of Muslim extremism and even giving extremists a place to live, health-care and money! In 2018 this idea of welcoming and promoting peoples who wish to exterminate us is growing. Behind it is – socialism!

It has destroyed British national identity; a deliberate activity meant to make the UK a puppet of the EU's and UN's formidable commands. It is not a surprise to find that Tony Blair, arch-Marxist, after ruining his own country, hoped one day to become President of the EU! Thankfully, this was a distant dream. Bill Clinton, another arch-Marxist, also tried to run the world by hoping to become Director-General of the UN, another Marxist organisation. Between them they can rule the earth with iron fists. However, Obama had his eye on lording it over everything, too – another socialist!

The only 'solution' modern Marxists have is depopulation. And they are not too fussy how this is done, including the UN 'Peacekeeper' troops who ravaged and destroyed the Congo, with the help of genocidal maniacs who call themselves 'presidents' of poor countries.

Marxists smile with satisfaction when poor farmers are denied modern machinery and markets. They want the world population to die down to about one fifth of what it is now. Of course, they do not want their own deaths, only those of people they think are surplus to requirement and unworthy to live.

It is from Marx that we get the idea of 'sustainability' (the passing-on of the earth to future generations, in a better state than we left it).

> "From the standpoint of a higher socio-economic formation, the private property of particular individuals in the earth will appear just as absurd as private property of one man in other men. Even an entire society, a nation, or all simultaneously existing societies taken together, are not owners of the earth. They are simply its possessors, its beneficiaries, and have to bequeath in an improved state to succeeding generations as boni patres familias (good heads of the household)." (26)

This is the basis for 'sustainability'. It is true that we merely live on the earth and do not own it. But, it is also true that science is trying to be boni patres familias. Many modern capitalists are turning to better ways of working, so Marx's complaints have no force today.

It was Clinton and Gore who forced non-action upon the USA, by barring production on mineral-filled land in the Yellowstone area.

Like the bad servant in the book of Matthew who buried his master's money, rather than risk trading, so these two men have 'buried' use of their country's land.

They neither moved forward nor backward, but sat on the fence, ruining the economy of the USA as a 'principle'. They did not even adhere to 'sustainability'. They just stopped-dead the possibility of improving lives and wealth. That is Marxism in action (or non-action)!

Yet, Marx said that the then new capitalism would need to reach a "higher synthesis" (the basis of dialectical materialism is to strike thesis against antithesis - counter-arguments - to produce a new synthesis that supposedly would be 'better').

So what was his problem? This 'higher synthesis' would require capitalism to deal with its 'human metabolism' in a 'rational way'. (27)

Is this not what modern capitalism is now doing? Yes, there are flaws and even deliberate avoidance, but in general this is what capitalism is enjoining, seeking ways of production that benefit both industries and human beings and the earth.

Marx's List Still Foundational

Marx and Engels spoke of soil nutrients. They also spoke of deforestation, desertification, climate change, elimination of deer from forests, pollution, industrial wastes, toxic contamination, recycling, exhaustion of coal mines, disease, overpopulation... (28). Charles Darwin contributed greatly to the work of Marx and Engels, though Darwin's claims would be challenged much later, highlighting, according to genuine modern scientists, many scientific problems in his rationale.

Marx and Engels had valid concerns at the time, but could not see beyond their own horizon, because their state of knowledge was fixed by what they then knew.

Today, Marxists have a very much better view of the world and greater knowledge, but they insist on going backwards to the lesser knowledge of Marx! Despite remaining flaws and deliberate capitalistic abuses, the general movement of the world is upwards, combining social conscience and wealth. So, Marxists have nothing in which to wallow... except in their addled minds.

It is the task of true science to falsify (find the flaws in) its own theories. In this way, newer theories (so long as they are logical and properly researched) take their place. But, Marxism dwells only on what is past, in an age we no longer know personally. That is why Marxism is dead, though it does not seem to know it yet.

Epicurus, the ancient philosopher, was Marx's hero, because he rejected all teleology and religious claims for existence. He also thought animals and human beings were the same. Thus, no matter how well received his theories are, they are just opinions and not fixed laws, as Marxists would have us believe.

Marx rejected ideas that did not mix with his own, so his intellectual endeavours were bound within walls, by pretending these other influences did not exist. This is why communism collapsed in Russia and can only be kept going elsewhere with brutal force, dictatorship, and the oppression of intellectual and religious movements.

It is rather like removing part of the equation, $A + B = C$. By taking out 'B' the equation does not work! But, it is used anyway, because those who prefer it are intellectually and morally blind.

> "He (Darwin) presented an account of the evolution of species that was dependent on no supernatural forces, no miraculous

agencies of any kind, but simply on nature's own workings." (29)

Because Marx saw this as a key component, his ideas threw out roots not into solid earth, but into the air. His trust in Darwinism was based not on proven science but on a scientific hypothesis that would later come under dramatic and deep scrutiny (not to be looked at in this volume, but see 'How the World was Made', 2018).

Let us imagine that Darwin did not exist and neither did evolution; would Marx still have written as he did? Yes, he would have. He had already begun his work before he learned of Darwin's theories. He was already predisposed towards materialism and antagonism towards religious thought.

As Marx said, "(Darwin's ideas are) the basis in natural science for our views." Thus, opinions were the basis for Marxism, not proven scientific facts. It does not matter if you are a Christian or not, Darwin's theories have never been proved to be true and are incapable of proof, because there is no way to replicate any evolutionary system or action within Darwinism. So, if Marxism is based on Darwinism, it stands to reason that Marx's ideas, too, are unproved and without foundation.

This has nothing to do with Christianity *per se*; it is all to do with the proper processes within science.

There is nothing wrong with following an opinion, so long as it is called an opinion and not fact.

When opinion and hypothesis replace actual facts and known truth, and is called 'law', we have a serious problem. And the problem is twice as bad if everyone is forced to obey what is hypothesis dressed up as law. If environmentalism wishes to be accepted as law, then it must show, through solidly logical scientific facts, that it is law and not mere hypothesis. And when paganism is added to hypothesis and both are added to enforcement, the problem   is   thrice   worse!

This is really the key to Marxist thinking, and it is prevalent today, as Marxist environmentalism attempts to cut religion (i.e. the 'B') from the equation, or any other theories, including scientific, that might upset its dictatorial course.

"The Soviet Union in the 1920's had the most developed ecological science in the world... that reaches down to the most advanced ecology of our day." Maybe so – but Soviet Russia died!

Stalin purged ecological components from Soviet leadership, because it seemed to resist socialist ideals by advocating a symbiosis with nature. After that, western Marxism divided its movement from nature. In other words, even that grand old man of evil, Stalin, could not see the link between Marxism and nature! Why, then, do modern Marxists see a link?

That is easy to answer: despite no link, Marxists will use environmentalism as a conduit through which they want to deliver noxious communism to the people, poisoning true intellectual, scientific and social advances.

The only exception is found, surprisingly, in Britain, which, to its cost, has always allowed extremists and those of dubious intellectual ideas, to thrive. "A strong tradition in Britain linked science, Darwin, Marx and dialectics." (30) Which is why the UK was one of the first modern western states to die of Marxian poison.

Marxist 'ecological science' so-called, developed mainly through this British connection, continuing to the present day. Even the supposed modern idea of species extinction is rooted in Marxism, through the work of Ray Lankester and Arthur Tansley.

Lankester was Huxley's protégé and the leading Darwinist of his day. His student, Tansley, was a close friend of Marx. Though Tansley was not himself a Marxist, Marx loved his work because he was a materialist, and offered to get his work published in Russian.

Lankester, alongside Marx, spoke with great urgency of the need to rid the Thames of its pollution, and rightly so. The same urgency is found today in all environmentalist claims... yet, the Thames is clean. Environmentalists now say there is no time to clean up the earth, but that was what Marx thought, too. The cleaning-up came, nevertheless! Tansley was the father of the 'ecosystem'. As we can see, then, Darwinist-Marxism is the foundation on which modern environmentalism is built. And their arguments are still buried under tons of nonsense.

Appearing at the same time as Stalin, Marxist biologist Lancelot Hogben challenged the ideas of Jan Christian Smuts. One of his objections was to Smut's 'racist eugenics'. Smuts was certainly not a nice man to know, but it is interesting that whilst this was an objection in the 1920's, modern environmentalist hard-liners want population to be cut far more aggressively. They relish wars that kill thousands, diseases that ravage poor Third World people, and refuse modern means to impoverished farmers to grow better crops. Anything that depopulates is fine by them! They became 'racist eugenicists' anyway!

Some environmentalists argue that Marx claimed men struggle against nature. This they have to claim to support their idea of nature being 'attacked' by humankind. Apart from finding ecological theories integrally in his works, Marx also fought Bruno Bauer who suggested "the antithesis in nature and history as though they were two separate things." (31)

So, Karl Marx is fundamental to modern environmentalism, as is Darwinism. Marxist theory is not fully-rounded or balanced and, when given its head, results in dictators, totalitarian regimes, oppression, loss of freedom, mass deaths, and decline. This is why we must beware. Anything Marxist is basically flawed and necrotic. It is intellectually stunted and causes mayhem. Why go back to Stalin for our way of living and working? Why allow politicians and others to inject us with Marxist ideology, when communism is dead and stinking?

Chapter 2

Fascism
and Environmentalism

Nice suit. Pity about the genocide.

Ted Kennedy was a typical fascist. He was one of those would-be rulers who are very vocal about everyone else complying with green culture and policies. But, when he was asked why he opposed a wind farm in Nantucket, he replied "That's where I sail!" Sorry, Mother Earth – Kennedy's boat comes first! Al Gore is just as hypocritical.

And where he sails is his personal waste-dump site... though he refers to Nantucket Sound as a "national (ecological) treasure", he had no problem commanding a crewman on his yacht to dump diesel into its waters. (32)

As the adage goes: "Do what I say, not what I do". Cynically, Kennedy, along with other wealthy rulers and users/abusers of people, only forced 'must do' orders on everyone else, not on himself. That, friends, is fascism!

One historian said

> "The goal of education in the past was to seek out the truth and to disseminate knowledge. The goal of education today is to promote a post-modernist version of diversity, multiculturalism, and political correctness, in spite of the truth. (Facts not wanted by these extremists) do not fit within the politically correct paradigm, and (are) therefore omitted from the curriculum and airwaves. (Many modern scientists, when finding evidence against their beliefs, label the evidence) 'anomalous', and (sweep it) under the rug. The textbooks and the media suppress the information. This is

especially true if the threatened paradigm has political ramifications." (33)

The truth about the environment is being suppressed. The media are part of this odd plan to deceive the public and will not publish anything against Goresque statements, even though most of them have been demolished or debunked.

In March 2008, the media proudly declared Tony Blair to be the champion of the environmentalist cause. But his opening statements are those that have already been debunked! He will go ahead anyway, just like any other Marxist-fascist. Truth does not matter and is not the issue. What matters is the creating of one world under 'Greens' (another name for fascists/Marxists), with all the oppression it means.

Strong-Arming

Marxism marches arm-in-arm with fascism (both are two sides of the same socialist coin). After all, oppressive ideology needs a strong arm to, well, strong-arm everyone.

> "... in addition to being America's most famous racialist, Madison Grant was one of the pioneers of American environmentalism. The eugenicist Henry Fairfield Osborn was the co-founder of the Save-the-Redwoods League. The

Third Reich was also a trail-blazer in conservationism." (34)

Note, though, that whilst Hitler conserved forests (except for those forests in other countries that his army blew up and burned) he shredded whole countries, bombed millions and gassed even more millions. And his neo-Nazi Green friends today can think of nothing better.

There is very little difference between Marxism and fascism! And no difference between fascism and the environmentalist idea of a 'solution' to depopulation.

Think my use of 'fascist' and 'Nazi' is unacceptable? Read this chapter and tell me there is no connection, if not a continuous link, between greens and Nazi socialism!

"The Nazis created nature preserves, contemplated sustainable forestry, curbed air pollution, and designed the autobahn highway network as a way of bringing Germans closer to nature. 'How Green Were the Nazis' is the first book to examine the ideology and practice of environmental protection in Nazi Germany. Environmentalists and conservationists in Germany welcomed the rise of the Nazi regime with open arms, for the most part, and hoped that it would bring about legal and institutional changes.... (The book) illuminates the ideological overlap between Nazi ideas and conservationist agendas. Moreover, this landmark book underscores that the 'green' policies of the Nazis were more than a mere episode or aberration in environmental history."

The book, written by a team of top-ranking historians, says the Nazi case raises questions about today's environmentalism, which they refer to as: "the 20th century's... most disruptive force." (35)

It is from these people-hating regimes and theories that we get organisations like Sea Shepherd, whose founder (and leader of the Sierra Club), Paul Watson, said: "Mankind is a virus and we need to 're-wild' the planet'." (36)

Watson is referred to as an 'eco-extremist'; similar views are sprinkled throughout the Gore-style movement literature and speeches (Gore is a friend of Watson). Watson "wants world population to drop below 1 billion." (37). But, his way of effecting that figure is dark and ominous.

He says "mankind is acting like a virus and is harming Mother Earth." He refers to mankind as a 'disease', "the AID's of the earth". His ideas may seem bizarre, but they are gaining ground, just as fascism gained ground in pre-Second World War Germany.

'How could it happen?' ask many. 'It must never happen again!' say even more. So, why do they vote for men like Gore and Obama?

In the UK a similar idea was invented around what are called 'Transition Towns'. Superficially, the idea is to get villages and towns to become independent of each other and to trade under a bartering system. I think you can see that this is a very good basis from which to go further into Watson-style theories:

> "No human community should be larger than 20,000 people, and separated from other communities by wilderness areas."

> "We need vast areas of the planet where human beings do not live at all, and where species are free to evolve without human interference."

> "Sea transportation should be by sail. The big clippers were the finest ships ever built and sufficient to our needs. Air transportation should be by solar-powered blimps…"

> "Essentially, Watson called for humans to return to primitive lifestyles. 'We need to

stop flying, stop driving cars, and jetting
around... The Mennonites survive
without cars and so can the rest of us."

Behind Watson's ideas is the concept of violent enforcement. Okay, try them out – but do not make it a legal requirement with penalties imposed for non-enactment! In his ideas we see not just something bizarre; we see the heavy shadows of Marx and Hitler. Indeed, a comment made about his ideas ran this way:

"This whole century, (a) long party of
massive breeding by the worthless, is
due to cheap energy. When cheap
energy goes, they go."

This is typical of the charming tirade that surrounds the environmentalism of the 21st century. Worthless? Does he mean like Jews were 'weeds' according to Nazi leaders? We all know what Nazis did to the Jews.

And note that making energy very expensive is a way to get rid of people... that is why energy is being pushed up in price in both the USA and the UK, where both governments are Marxist-fascist socialist (even Tories/conservatives). You are suffering because you voted them in! And the EU is hiking up energy prices, too.

Hitler's Mean Green Machine

A number of books have tried to decipher whether or not Hitler was a Green. (38)   The Editors of 'Blood and Soil' say there is "no linear relationship" between "today's Greens and yesterday's environmentalists". (39). But, as you can see in chapter one, the same denial was made about Marxism!

The editors, like so many who want to sanitise a disgusting movement, say we must have a "value-free" analysis of Nazi environmental policies, so that we may have a 'balanced' view: "to miss the positive features of National Socialism is to miss why it appealed to so many people" (40)

This kind of statement is dangerous, if not outrageous, because few ordinary people understand how to separate beliefs and statements logically, without emotion.

Therefore, there is the very real possibility of 'environmentalists' (that is, rank-and-file members or sympathisers, with no idea about what environmentalism really means) just accepting Nazi Green policies as they stand. One cannot be 'value-free' when discussing something as horrific as a movement that obliterated millions of people at whim... and which is now rising again, speaking of 'depopulation'!

"The authors stress the sometimes uneasy relationship between Nazi ideology, and policy, and 'green' ideas." I bet there is! "In fact, while in many cases there were opportunities for cooperation, in others, such as the Nazi rearmament policy, there were unbearable frictions". (41)

"For some green-leaning Nazis, however, that was acceptable. For them the war and destruction were necessary evils, since they would bring about a new order that would finally allow the establishment of a better and greener Germany." (42)

Another reviewer adds:

"This book promises to be one of the most important reference books for understanding the links between violence and green thought, and the need to look at environmentalism as a value-laden enterprise." (43)

Alarmingly, most Greens at the time wholeheartedly adopted the National Socialist Party as their own (chapter one), because they thought they would then see progress in terms of environmentalism. But, while waiting to see the results of Green policies, in particular the Reichsnaturschutzgesetz (RNG; Reich Nature Protection Law), these undiscerning people accepted all that went with Nazi policy, including its thuggery, enforced acceptance and racism.

When Nazis proved to be less than eager to make a success of the environment, however, the Greens slowly disaffected themselves. Perhaps Hitler's mind was on far more than flowers and soil!

The environmental concerns of the philosopher, Martin Heidegger, were once thought to be spawned by Nazi ideals. Later research suggests he did not enjoin environmentalism until he became disappointed with socialism's lack of movement towards it. Even so, the link is there.

Chapter nine of the book portrays a different reference, looking at the violence in Nazi environmentalism, quoting the way western Poland was annexed as a prelude to invasion of the whole country. German landscape architects enthusiastically built Nazi Green ideas into a wide-scale environmental upheaval of eastern Poland, enjoying the total freedom to do what they liked. Not because east Poland was a waste area, which it was, but because it was to be made beautiful for the coming Germans who would push Polish families out of their homes.

Very clearly, then, Nazi environmentalism was an expression of its socialist fascist ideals, which were pressed onto whole countries to get rid of their national characteristics. (Denationalisation is happening now in the UK, thanks to its former PM, Tony Blair, under orders from the EU). "By obliterating the visual structures of Polish culture, they participated in their own unique way in the implementation of the 'final solution'." (44)

Look at the aims of Greens today, especially those who are hardened activists, because it is they who lead the movement. You will find mirrored ideals concerning the 'final solution'. Later, you will read quotes by world-known Greens that should send chills down your spine. Like the Nazis, they want a 'final solution' to population!

Do not be fooled into thinking environmentalists only have an interest in wild gardens and organic food! They want total control over every aspect of your life, with severe penalties if you do not comply. And they want to see swift depopulation, especially amongst the poorer nations. Ask yourself: how do you get 'swift depopulation', except through death?

Now jump forward a few decades and see fascism today. To quote a blog by Leo Oshkosh: "... the liberal wing of the (US) Democrat Party believes in many of the same principles adopted by Adolph Hitler". (45) He is being modest, because the Democrats portray a virulent form of Marxism coupled with fascist techniques.

The blogger then mentions that Jonah Goldberg was writing a book called 'Liberal Fascism: The Totalitarian Temptation from Mussolini to Hillary Clinton.' It was published in 2006. It is fact that both the Clintons are Marxist and tend towards a hardened hatred for people. In her presidential campaign Clinton proved herself to be a nasty piece of work.

Oshkosh lists several traits of liberals that coincide with Hitler's:

1.   He was Vegetarian: Most liberals are vegetarian and, usually, animal-rights players. Hitler once said that the future world will be populated by vegetarians. (46)

2.   Hitler and Himmler Play with the Environment. In December 1942, Himmler released a decree already referred to above: "On the treatment of the Land in the Eastern territories". He was talking about the annexing of eastern Poland, and said:

> "The peasant of our racial stock has always carefully endeavoured to increase the natural powers of the soil, plants and animals, and to preserve the balance of the whole of nature. For him, respect for divine creation is the measure of all culture. If, therefore, the new Lebensräume (living spaces) are to

become a homeland for our settlers, the planned arrangement of the landscape to keep it close to nature is a decisive prerequisite. It is one of the bases for fortifying the German Volk." (47)

3.   Hitler hated Capitalism. He scorned capitalists publicly, because he said they controlled the masses with wages and goods. Not very different from Marx. In the 1940'e he forced capitalists to distribute goods equally (more equally to Nazi leaders, of course, and he and his cronies amassed unbelievable hoards of gold and art works for themselves). The blogger thinks Hillary Clinton (and most Democrats, for that matter) would love to be part of Hitler's regime, using higher taxes and the creation of countless government programs to spread wealth not only to citizens, but also to illegal immigrants in their millions. (Illegal immigration is a good way to destabilise a country and destroy its national and cultural roots. Recognise it as both Marxist and fascist? See it in the USA under Obama?).

Can We Legitimately Refer to Hitler?
It is true that both sides of a heated argument will, at some time, draw on the 'Hitler Card', in an effort to disparage the other person. (fallacyfiles.org). On many occasions the card is invalid. But, is it invalid in the case of environmentalism? No, it is not, for what   we   are   doing   is   showing   that

environmentalism finds its roots and methods in both Marxism and fascism.

The reason for highlighting the wickedness of both movements is that present day environmentalists are showing the same characteristics: particularly, the forcing of others to act as they want them to, the spreading of deceptive propaganda, the character-assassination of anyone who teaches otherwise, the reliance on false science, the stranglehold they wish to place on everyone by way of taxation and energy controls, and the ominous call for a 'solution' to population (a call that is growing, but not including those who call for it!).

This is no mere 'guilt by association' exercise, but a 'familial guilt by direct link with fascism'! It is not a way of derailing Green arguments – it is based on fact, as has been shown.

It is interesting that Marxists are considered leftist and fascists are rightists. It is interesting because modern environmentalists are both. They have joined forces pseudo-intellectually to bring about the desired objective. The objective is not very nice, and is the plaything of the rich and powerful. The poor, those who do not lead, and the plain ignorant, are just pawns.

It might be argued that the concerns of people today are those of informed thinkers who truly believe there is environmental danger (and incudes the climate change lobby, which bases itself on pseudo-science). This is to misrepresent simple group dynamics. The expression of concern by the public is not down to actual understanding, but down to expert manipulation of the public by propagandists.

The British government is 'spreading the word' via schools and in continuous media ads. The message is very simple and factually wrong, but, because it is said often enough, it is believed by the masses... a very Hitlerian tactic!

People have been scared witless, because they have been told they will be drowned by serious sea level rises within the next ten years, so "we must act fast". Tony Blair spread the same word on behalf of the EU... and his motives stank to high heaven, as does the fake science behind the claims.

Britons - especially children, which is appalling - have been scared into thinking climate change will bring about massive immigration by hungry people who will use war to get food. And it is believed, even though there is no science or any other proof behind it. Even today the mass migration of Muslims to the West has nothing to do with food, but with being paid to invade by Islamic terrorists, so the West will be destabilised by rampant Islam.

It is done by presenting a fearsome future and by eradicating and silencing opposing views, though there are no real facts. In this the media are almost 100% complicit. The aim is to control the people by taxation and law and a tight leash on energy, with a more punitive regime to continue.

That, friends, is fascism! And that is why I link it with Nazi Germany and Hitler. G. Gordon Liddy, radio host and former Nixon official, openly admits that his views on environmentalism are the direct result of being inspired by Adolph Hitler in his youth. Liddy said:

> "Environmentalism is a form of pagan
> fundamentalism. These green wackos
> are fanatics like al-Quaida... the
> environmentalist believes human beings
> cause global warming. (Like Muslim
> extremists) they both want to wreak
> havoc because of their mad beliefs.
> What's the difference?"

And he refers to Greens as being "environmentally ill" and "multilateralist UN one-world government worshippers and other politically correct castrati. Why?" (48)

Though Liddy speaks the truth on environmentalism, his love of Hitler is not too hard to fathom. As a boy he lived in Hoboken, New Jersey, a town full of ethnic Germans who idolised

Hitler. In school, the nuns made him salute the Stars and Stripes Nazi-style. And he had a German nanny who taught him that Hitler was a wonderful leader.

In the same way, people today are being brainwashed by governments and environmentalists into believing death is coming in the next mail and everyone is doomed. There is no science behind it, just stuff and nonsense.

But, that is what social dynamics is like: we believe it because everyone else believes it, and they believe it because scientists say so, so it must be true! This is the believism of the school-yard, not of responsible adults!

Fascist Ecology

> "We recognise that separating humanity from nature, from the whole of life, leads to humankind's own destruction and to the death of nations. Only through a re-integration of humanity into the whole of nature can our people be made stronger. That is the fundamental point of the biological tasks of our age.
>
> Humankind alone is no longer the focus of thought, but rather life as a whole… This striving towards connectedness with the totality of life, with nature itself, a

nature into which we are born, this is the deepest meaning and the true essence of National Socialist thought."(49)

There are fascists who are concerned with the environment, and there are environmentalists who have fascist traits. They are different, but we cannot thereby dismiss the fascist principles in either.

To repeat - fascism insists on obedience to its regime, whether or not it is believed; it destroys and suppresses freedom of speech and thought (which is oppression); it penalises all who object or who oppose; it enforces through control of public information and resources.

Though this ought to remain a purely political fascist domain, it is not; it is found aplenty in environmentalism and throughout modern societies. The Transition Town* idea is replete with both Marxism and fascism, using soft words! In the UK, Marxists have openly claimed the movement to be its own, because, as they say, it is an excellent medium to further their cause. (*For more on this see the 2009 book).

Care of Forests, but not Humans
Even so, there appears to be a refusal to acknowledge the link between fascism and ecology, or to examine the "ideological overlap between nature conservatism and National Socialism". (50)

The Hitlerian idea of Green policies came from earlier German thought. For example, Ernst Moritz Arndt. In an 1815 article, 'On the Care and Conservation of Forests', he passionately spoke against deforestation, over-exploitation of woodlands and soil, and the economic uses of it, in extreme and unreasonable terms:

> "When one sees nature in a necessary
> connectedness and interrelationship,
> then all things are equally important –
> shrub, worm, plant, human, stone,
> nothing first or last, but all one single
> unity." (51)

Arndt was also an ardent and fanatical nationalist, and always taught everything in terms of German soil and German people.

> "At the very outset of the 19th century the
> deadly connection between love of land
> and militant racist nationalism was firmly
> set in place." (53)

Now, militant nationalism has been replaced by global militancy fighting for a global 'solution'. Hitler's dream is at last finding its feet and running hard!

Romantic Thuggery
Wilhelm Heinrich Riehl, a student of Arndt, "developed (his) sinister tradition", his ideas

"presaging certain tendencies in recent environmental activism; his 1853 essay, 'Field and Forest', ended with a call to fight for the 'rights of wilderness'." (53)

The only difference between this call and today's call, is that today environmentalists are hysterically demanding global, not just national, control.

Riehl loathed industrialisation and urbanisation. He glorified rural peasant values (with antisemiticism thrown in for good measure) and hated modernisation, thus earning the title "founder of agrarian romanticism and anti-urbanism" (54). Of course, only those with money and power can be romantic about crippling 'peasant values', because they do not have to live as peasants! Similar romanticised ideals are found in abundance in the Transition Town theory.

This nationalist obsession with ecological 'purity' (only available through the Arian race) led to the Völkisch movement that combined 'ethnocentric populism with nature mysticism' (55). Modern environmentalism is very similar in its approach; note the strong paganistic rituals and beliefs emerging and underpinning Green mantras.

Völkisch followers wanted a return to the land and simplicity. Nothing wrong with that… if that is what each person wants. The difference today is that Greens want such a move backwards to be a command upon all people, whether or not they want it. It is a worrying element in the Transition Town idea, now gaining ground in the UK and USA.

It is this enforcing that indicates a fascist root. And it is a movement that wants to get back to a romantic notion of the land, not shared by those who are actually shackled to it because of low status and position in life!

It would be wonderful to have an idealised countrified life. But, it is impossible for most people. The irony of it is that it has been made impossible by crippling national laws and Marxist-fascist governments who refuse to allow citizens to just get on with life as they see fit!

"The mystical effusiveness of this perverted utopianism was matched by its political vulgarity" (55). This is exactly what we are now seeing in Green politics! No well-honed arguments coupled with scientific facts, just paganistic statements coupled with bad science, both foisted on us by ignorant governments and greens.

The Völkisch movement "aspired to reconstruct the society that was sanctioned by history, rooted in nature, and in communion with the cosmic life spirit." (56)

But note – the enthusiasm for all this ecological wonder was not really about love of nature, but was an expression of hatred for Jews which would soon murder Jews in their millions.

> "The Germans were in search of a
> mysterious wholeness that would restore
> them to primeval happiness, destroying
> the hostile milieu of urban industrial
> civilisation that the Jewish conspiracy
> had foisted on them." (57)

Modern green hatred is not just reserved for Jews; it is aimed at Christians and Muslims as well; the key being any religion that worships a personal God (as exemplified by Al Gore's book). In other words, today's Green movement mirrors the hate-ridden Völkisch movement that predated and gave birth to the National Socialist ideology of Hitler.

Throughout, the emphasis is on a romantic idea of nature, not a practical or actual idea. It is rather like imposing the 'ideals' of an animated cartoon film on real society!

> "The emergence of modern ecology
> forged the final link in the fateful chain
> which bound together aggressive
> nationalism, mystically charged racism,
> and environmentalist predilections. In
> 1867, the German zoologist, Ernst
> Haeckel, coined the term 'ecology'... (he
> was) also the chief populariser of Darwin
> and evolutionary theory... (developing) a

peculiar sort of social Darwinist philosophy he called 'monism'. The German Monist League he founded combined scientifically-based ecological holism with völkisch social views. Haeckel believed in Nordic racial superiority, strenuously opposed race mixing and enthusiastically supported racial eugenics." (58)

"Haeckel contributed to that special variety of German thought which served as the seed bed for National Socialism... one of Germany's major ideologists for racism, nationalism and imperialism." (59). Later he joined the Thule Society, a radical group that helped establish the Nazi movement. (60)

"The pioneer of scientific ecology (Haeckel), along with his disciples... profoundly shaped the thinking of subsequent generations of environmentalists, by embedding concern for the natural world in a tightly woven web of regressive social themes. From its very beginnings, then, ecology was bound up in an intensely reactionary political framework." (61)

Al Gore has been spinning the derogatory title 'deniers', which he uses with force against anyone who does not accept his very unscientific views. He has brainwashed the media into ignoring opposition to his Marxist beliefs and fascist enforcements. (Obama did the same). Yet, his attitude and methods were spawned by both Marx and Hitler, encouraged by the ultra-right Watson of the Sierra Club.

> "For the Monists, perhaps the most pernicious feature of European bourgeois civilisation was the inflated importance which it attached to the idea of man in general, to his existence, and to his talents, and to the belief that through his unique rational faculties, man could essentially recreate the world and bring about a universally more harmonious and ethically-just social order. (Humankind was) an insignificant creature when viewed as part of, and measured against, the vastness of the cosmos and the overwhelming forces of nature." (62)

There is an innate absurdity, if not an insanity, in all this. Because the cosmos is bigger than human beings, human beings should just sit around and let 'nature' kill him! For Greens, Man is nothing compared to nature, and evolution is all there is.

This is nonsense. Man has a brain and the ability to remake earth's component parts into useful things. From earth he can grow food. From trees and ores he can make objects. From the sea he can bring fish for food.

But, environmentalists do not like this. They want the earth, with all its treasures, to be left as it is, and if anyone dares to trifle with it, they must be punished. In this way, Greens place the inanimate higher than the animate, the unthinking above the thinker; claiming the iron effigy of their god is real. Better to starve to death than use up bits of 'mother earth'! (More precisely, others can starve, while they survive!).

A founder of the Monist movement, Raoul Francé, "is acclaimed by modern ecofascists as a 'pioneer of the ecology movement.'. " (63) How can anyone, then, say fascism has no bearing on modern environmentalism, when ecologists themselves honour and follow early fascists, have the same beliefs, and use the same fascist techniques? If it quacks like a duck, waddles like a duck, looks like a duck, well... In keeping with Darwinism, fascists view social behaviour through biological lenses.

> "During the turbulent period surrounding World War One, the mixture of ethnocentric fanaticism, regressive rejection of modernity and genuine environmental concern, proved to be a very potent potion indeed." (64)

Youth as a Tool of Insanity

During the Weimar era (between the wars), especially when Hitler came to power, youth was manipulated and used to spread the vile fascist dogma. Today, rampant homosexuals and Islamists do exactly the same thing – targeting children and youth, because they are malleable.

Al Gore and governments are brainwashing youth in schools with the same dogma, knowing the truism used extensively by Jesuits, that if you capture people young enough, they can be made to do anything you teach them to do.

That is, indoctrination. Schools and universities are no longer places of true learning, but dark caves illuminated only by what the person with the torch tells you.

The Weimar age, with its brainwashed youth, easily spread fascist ideals throughout society. Young people are famed for seeing everything in black and white, thinking 'their' ideas are their own, and getting excited about 'their' discoveries!

They are too inexperienced to realise 'their' ideas are placed there by clever and unscrupulous adults who are guiding their every thought towards a hidden agenda. Greens today are scaring children witless, making them think the sea will drown them if CO_2 does not smother them! And so children are at the forefront of Green public statements. It is dependent on Greens keeping children from having two sides of the argument. But, children could not be scared so much unless their own parents allowed it to happen! What does this tell us about parents today?

> "The youth movement (Wandervögel)
> was a hodge-podge of counter-cultural
> elements, blending... romanticism,
> Eastern philosophies, nature mysticism,
> hostility to reason... Their back-to-the-
> land emphasis spurred a passionate
> sensitivity to the natural world and the
> 'damage' it suffered... most of the
> Wandervögel were eventually absorbed
> by the Nazis. This shift from nature
> worship to Führer worship is worth
> examining." (65)

The romanticism of the movement led it to the:

> "unpolitical zealotry of fascism... The
> youth movement did not simply fail in its
> chosen form of protest (romanticism), it
> was actively realigned when its members
> went over to the Nazis by the thousands."
> (66)

A movement that does not recognise its own roots, and experiences what it perceives to be political estrangement "can, in times of crisis, yield barbaric results." (spunk.org). That is why today's Marxist-fascists again resort to eugenics; hence abundant abortions. To ignore history is to ruin your future.

Can you not see that the historic fascism of environmentalism will result in the worst possible scenarios? Do you think neo-fascism, if given its head, cannot seek the same 'barbaric results' today?
If the marks of fascism are constant throughout history, why do you trust its methods and aims... unless its aims are also your own?

Man and Earth
In 1913, the Wandervögel gathered in Meissner. In that gathering the philosopher, Ludwig Klages, presented his now famous essay, 'Man and Earth'. Already influential with German youth on ecological matters, this paper made him supreme. The essay is counted to be "one of the very greatest manifestoes of the radical ecopacifist movement in Germany". (67)

More significantly, it is "also a classic example of the seductive terminology of reactionary ecology" referred to today by environmentalists. (68)

The paper is like a modern Green wailing exercise, without true science behind it: It spoke of acceleration of extinction of many species; disturbance of 'global ecosystem balance' (how could they assess anything 'global' at that time?); deforestation; destruction of aboriginal peoples and wild habitats; urban sprawl, and the alienation of people from nature.

Interestingly,

> "it disparaged Christianity, capitalism, economic utilitarianism, hyperconsumption and the idealogy of progress. It even condemned the environmental destructiveness of rampant tourism and the slaughter of whales, and displayed a clear recognition of the planet as an ecological totality."
> (69)

In this we see the obvious link between earlier ecology and modern environmentalism. In it we also see the finger that prodded Al Gore and his Green associates, who blame Christianity, capitalism, et al. Same arguments, same bad science, and same irrelevant conclusions. And Klages, too, was "a venomous antisemite".

Hatred and fascism go hand in hand. Environmentalism, as a grandchild of these people, is a seedbed for hatred, carrying with it the same genes of destruction.

(Whether or not you are a Christian, or other religious person, the major point here is that a large section of the people are ignored and demonised by another section, just to continue a hateful and ignorant regime; this is how all socialists work. And when Christians are finally put into an intellectual – and maybe actual – pit, the Greens will go after someone else they think is 'responsible' for ruining 'mother earth'. Stalin did the same thing. Beware!).

Klage manufactured a new meaning for 'Geist' (Ed. denunciation of rational reason itself):

"Such a wholesale indictment of reason cannot help but have savage political implications. It forecloses any chance of rationally reconstructing society's relationship with nature, and justifies the most brutal authoritarianism". (70)

But, the lessons of Klage's life and work have been hard for ecologists to learn. "In 1980, 'Man and Earth' was republished as an esteemed and seminal treatise to accompany the birth of the German Greens." (71)

Does this not frighten you? And does it not clearly show that fascism is not just alive and kicking, but gaining new impetus from the republication of antisemitic, hateful literature from the past? The link between then and now, early ecologists and modern environmentalists, is very obvious.

Heidegger and Ecology
Philosopher Martin Heidegger "helped bridge fascism and environmentalism" and hotly defamed

modern technology. Therefore, he is "often celebrated as a precursor of ecological thinking... Contemporary deep ecologists have elevated (him) to their pantheon of eco-heroes." (72)

"Heidegger's critique... his call for humanity to learn to 'let things be', his notion that humanity is involved in a 'play' or 'dance' with earth, sky, and gods, his meditation on the possibility of an authentic mode of 'dwelling' on the earth, his complaint that industrial technology is laying waste to the earth, his emphasis on the importance of local place and 'homeland', his claim that humanity should guard and preserve things, instead of dominating them – all these aspects of Heidegger's thought help to support the claim that he is a major deep ecological theorist." (73)

For environmentalists to pretend there is no continuous link between these pre-World War Two figures, whose ideas brought Nazi thinking into being, and modern environmentalism, is absurd. It is clear as day! It is also clear in the Transition Town-style idea led by zealots without true knowledge. Hidden in their seemingly kind motives there is the spectre of genocide, often referred to as 'population control'.

It is true that Zimmerman has since thought again and urged readers not to adopt Heidegger as their hero. But, how many will listen now? Very few, because the book suits their already fascist mindset. Heidegger was an active member of the Nazi party and adored Hitler. Enough said!

Heidegger lived another thirty years after the fall of Nazi Germany, yet he never once renounced his anti-Semitism, his fascist ideology, or Nazi crimes.

"His work, whatever its philosophical merits, stands today as a signal admonition about the political uses of anti-humanism in ecological guise." (74)

But who will similarly rebuke fake environmentalists, Greens, our modern fascists? And who will rebuke and depose the arch-Marxist-fascist of our day, Al Gore? (We can now add Tony Blair and many others to that list).

When men continue to push a movement that has been soundly debunked, this is evidence that the movement is not the aim, but its political power and benefit to the one pushing it, is the true goal.

"The Nazi movement's incorporation of environmentalist themes was a crucial factor in its rise to popularity and state power." (75)

In the same way, rampant power-hungry reactionaries are using environmentalism to further a political regime of destructive force, not just Germanically, but globally. Both the EU and UN are pawns of this regime (They are willing servants of very wealthy socialist dictators). You must seriously and urgently ask "Why?"

Historian, Robert Pois, gives a stark warning:

> "The National Socialist 'religion of nature', was a volatile admixture of primeval teutonic nature mysticism, pseudo-scientific ecology, irrationalist anti-humanism, and a mythology of racial salvation through a return to the land. Its predominant themes were 'natural order', organicist holism, and denigration of humanity." (76)

Note the denigration of humanity, the irrationality and pseudo-science. This is a strong theme throughout environmentalism and is boldly being preached by leaders antagonistic to people. It finds itself embedded in mass abortion, homosexual addiction, transgender delusion, and many more 'movements' that lead to the death of mankind or its fast ruin.

These leaders who push these ideologies want people dead, not alive, because people, they say, are ruining the earth. More specifically, they are stopping those hidden wealthy totalitarians (Soros is a prime example) gaining world power.

You should oppose these teachings, because we saw what happened when they took hold in Nazi Germany, and in many countries since, including Hussein's Iraq, communist China, Vietnam, and African despotic regimes.

In essence, fake environmentalists in the West want to stay alive, at the expense of millions dead in the east and 'Third World' countries.

> "Throughout the writings, not only of Hitler, but of many Nazi ideologues, one can discern a fundamental deprecation of humans vis-à-vis nature, and, as a logical corollary to this, an attack upon human efforts to master nature... man is a link in the living chain of nature, just as any other organism." (77)

We see in this at least one reason why modern environmentalists and earlier ecologists, decried Christianity...

Christians (not the lame ducks found today, who support Al Gore because he pretends to be one of them) are given the command to populate the earth and to master nature, not to allow nature to dictate its own terms. Greens hate this because they hate God, and so they attack Christians who speak the truth.

Environmentalism wants the flower of humanity to be choked by weeds, because the Green gardener refuses to believe some things are weeds and others are things of greater beauty and worth. Green roots (socialism) are dead, but still revered.

Modern Greens follow Hitler in their hatred for people. Gore guru, James Hansen, said:

> "Some of this noise (i.e. opposition to global warming claims) won't stop until some of these scientists are dead." (78)

Charles Wursta, chief scientist for the Environmental Defense Fund, in responding to the claim that banning DDT could put millions of lives at risk, said, with great empathy for poor people:

> "This is as good a way to get rid of them as any". (79)

Think these are random quotes by Greens? Lamont Cole said:

> "To feed a starving child is to exacerbate the world population problem." (80)

Greenpeace co-founder and Sierra Club leader, Paul Watson, is typically fascist:

> "I got the impression that instead of going

out to shoot birds, I should go out and
shoot the kids who shoot the birds." (81)

You might think it is the kind of thing anyone would
say... but he really means it! His 'Club' wants mass
deaths so as to 'save the earth'.

Now, if he said "shoot the pro-Green politicians" I
could understand that, because they are so full of
hot air about global warming! (That was a joke, not
my actual belief or a policy!)

Since saying that, Watson has matured into a fine
specimen of hate, combining both Marxism and
fascism, and leading Greens down a path that, in
earlier times, led to genocide. It still does.

To ignore such hate and to say modern
environmentalism has no connection to Hitler or
fascism, is a negation of the facts and the obvious
marks of fascism found in today's liars and frauds,
who repeatedly cry loudly that we are doomed
unless we pay higher taxes to stop climate change!
Eh? This is not the first time that "The earth will die
in ten years" claim has been made! But, the earth is
still here!

Fascism did not just suddenly appear; it had to be
nurtured and whipped into a frenzy. We are seeing
this today. In particular, we see it aplenty in the
USA where anarchy and socialism are literally

screaming in the streets. It is also seen in the rapid Islamisation of Europe. It is all the same socialism.

Hitler often impressed upon his captive audiences the "helplessness of humankind in the face of nature's everlasting law" (82). Not bad for a housepainter! He was not a physical scientist, yet he imposed his will upon the people with odd scientific 'laws' he believed existed.

This is just like people today, including Al Gore, who have no idea if what they are being told is true, but believe it anyway. And that includes the hatred shown for humanity and progress. Thus, it is political theory and not scientific law that matters. Science, though the bad variety, is useful as a lever.

> "When people attempt to rebel against the iron logic of nature, they come into conflict with the very same principles to which they owe their existence as human beings. Their actions against nature must lead to their own downfall."

Guess who said that? Al Gore? One of his gurus? It certainly sounds like them! But, it was said by Hitler in his book, Mein Kampf. (83)

Are you not disturbed to read those words by Hitler, when similar words can be heard constantly coming from mouths of hatred, such as Al Gore, the

Clintons, Tony Blair, and Obama, and foolish world leaders who copy them?

The Nazi Machine

The Nazi machine indoctrinated German youth (as it does today in US and UK schools) to believe that they must strive constantly, as a matter of "civic importance", to coordinate all parts of society and life "for the benefit of the one and superior task of life". (84) That is, Hitler youth were taught that men are subservient to nature and so all their actions must serve that nature... by force if needs be, or by force as a matter of policy. And it is the same hatred that attacks Christians today.

These ideas were used to rationalise the fascist regime of the Third Reich, as well as the annexing of Poland and other countries, and altering their landscape to suit Germanic pride. Today, this expansionism and fascism is found in excessive doses in environmentalism as a tool of EU and UN dreams of dictatorship.

The idea also "provided the link between environmental purity and racial purity". (85) Look at modern environmental quotes (some are above) that very obviously and hatefully muse about Third World deaths as happy events, because they reduce population figures without any effort. Some quotes show that withholding help to starving peoples is a good thing. One way is to stop transport of goods and food from the Third World for 'Green' reasons, thus denying trade for poor peoples, which, of course, leads to their demise.

The next step must be deliberate genocide (a UN specialty). Speculation? No, just observation. These people hate everyone but themselves and are more than willing to 'press the button'. This is because they follow Uncle Hitler and his fearsome brand of fascism, and it is a logical progression:

> "Two central themes of biology education follow (according to the Nazis) from the holistic perspective: nature protection and eugenics. If one views nature as a unified whole, students will automatically develop a sense for ecology and environmental conservation. At the same time, the nature protection concept will direct attention to the urbanised and 'overcivilised' modern human race." (86)

Automatically, then, this kind of approach to nature could not be assumed by Christianity, which looks upon humanity as the highest form of being on earth, animals as lower, and the rest, from the earth and earth itself, as subordinate to humankind, to be used as mankind sees fit. The fascist regime is tthe antithesis. So, Christianity must go. (This includes any other group worshipping a personal God).

Therefore, modern environmentalism, which is internally opposed to Christianity et al, and based on fascism, has no other option but to have the same hatred as was found in Nazis. Which brings up a conundrum that was also found in Nazi Germany... the most supportive group of Hitler's ideas in pre-Second World War Germany were – duped Jews. But, this did not stop Hitler destroying them when they had foolishly brought him to power!

Today, many churches and pseudo-Christians are supporting Al Gore, even though he has openly denounced Christianity. It does not make sense, and will prove to be the downfall of all 'Christians' who support him (and who have no idea that, according to the Bible, the earth belongs to mankind, who may use it as they wish). He is a Marxist-fascist and must follow his fascist roots. And so must they. (See my short book on the so-called 'Green Bible').

Pagan Ideology
Himmler, along with others who loved the romantic concept of nature, was a neo-pagan, hence, like Hitler, he practised witchcraft. And what do we see underpinning modern environmentalism? Yes, paganism, most pronounced in the Transitional Towns-style movement.

The paganism in Hitler's time is exactly like today's romantic versions, and reflect Nazi practices "which are today conventionally associated with ecological attitudes." (87) Both Himmler and Hitler were strict vegetarians, animal lovers, into nature mysticism and homeopathy, and opposed to vivisection and cruelty to animals. (88)

Is it not strange? They had all those beliefs and yet it was acceptable to vivisect human beings because they were Jews, and it was acceptable to be cruel to people of any type? Hitler often spoke of renewable energy sources as alternatives to coal; he specially commended "water, winds and tides" as the sources of energy of the future (89). Sound familiar?

As others have said, this comprises all the usual statements made by "classical ecofascist ideology" (55). So, as I have already said, if it looks like a duck, sounds like a duck and waddles like a duck... maybe it is a duck after all!

Modern environmentalists are merely repeating the views and beliefs held by the master fascist, Hitler. Therefore, they are fascists.

And, as was said of Nazi Germany, "Clearly, the affinities between environmentalism and National Socialism ran deep." (90) Environmentalism and fascism, then, come from the same wicked womb.

In 1930, it was top fascist, Richard Walther Darré, who said "The unity of blood and soil must be restored". (91) The 'blood' referred to the race or nation; 'soil' referred to land and the natural environment.

Because Jews were always wandering (precisely because they have always been harassed and attacked by groups like the German fascists) they were deemed not to have any kind of relationship with the earth, so were like weeds ready to be burned.

'Blood and soil' became a central statement of Nazi policy. And Darré, as Reich Peasant Leader and Minister of Agriculture, made sure ecology was given supreme support. (92)

What is emerging here? We are seeing that environmentalism was not the main issue. Ecology was only the front for political change, and used to horrendous effect by ruthless leaders. This is the current situation, too, because fascism, does not change; it only changes its names to hide the truth.

Al Gore, Tony Blair, Clinton, Obama, the EU and the UN – all want to change societies worldwide so that they comply with their inordinate and selfish desires for power and control, and, of course, a lot of cash. They are using environmentalism as a vehicle to ram-raid our senses and bring in their own dictatorial aims. Ignore it at your peril!

Darré used Green policies to justify violent Nazi colonisation of Poland, "as the very basis of the Third Reich's agricultural policy. Even in its most productive phases, these precepts remained emblematic of Nazi doctrine." (93) Nazi - fascist -

environmentalism - Gore/Blair/Obama - UN, EU; Hitler is alive and well.

One writer refers to Darré as the "father of the Greens". (94) This title was awarded to a man who called Jews 'weeds' and was a disgusting type of racist, fully in agreement with 'final solution' policies (if not the actual author).

> "The ecological aspects of his thought cannot... be separated from their thoroughly Nazi framework.... He represents the baleful spectre of ecofascism in power." (95)

This same ideal is now jostling to be put into power again by modern environmentalists... read the great number of pro-Green remarks made on websites and in blogs. If you do not recognise, and oppose, their strident virulence, then you will soon become a willing slave... or dead, if you live in the Third World.

Neo-Fascism, Son of Historical Fascism

The problems I am highlighting are not about using best possible farming methods or reducing waste, etc. All ages should use resources wisely, with an eye on what is best practice. The issue is to do with the 'genetic' link between fascism and environmentalism.

The combination is a death-blow to genuine freedom and ability to choose. Time and again, references in historical narratives to German ecology show the symbiosis of fascism and environmentalism.

Another "fanatical ecologist" (96) was Alwin Seifert, chief adviser to Fritz Todt. He was the Reich Advocate for the Landscape and was nicknamed 'Mr. Mother Earth'. It is amusing because this is what I call Al Gore!

Seifert wanted a reverse of progress, from technology back to nature. (97) This is the call of the wild amongst today's environmentalists and Transition Town-style groups.

All of Seifert's ecological wishes are found in modern Green policies. The words are the same! And the methods are coming dangerously close to the same, too! Todt and Seifert tried to bring in a 'Reich Law for the Protection of Mother Earth'. Is this not what modern manic environmentalism is seeking, through the Kyoto and Bali etc., agreements? German ministries at the time accepted the notion – except the ministry responsible for coal mining. Even this is similar to today's situation.

Reich Chancellor Rudolph Hess made sure these ideas stayed at the very top of the Nazi pile and it was through his law-making power that Greens stayed there, with complete confidence in full fascist sympathy. They fed each other.

Hess was very strict in adhering to a 'natural' lifestyle, down to a biodynamic diet that was more stern than Hitler's. (A miserable existence seems to be a constituent part of being a fascist!).

Though close to Darré, Hess was even more fanatical about organic farming, and employed numerous specialists to ensure Seifert's ideas were spread around Germany and annexed lands. (98) Hess brought many Green laws into being, swiftly and with force as needed; something akin to the speed and determination seen today, when Greens easily use force and hate to get their own way.

Like Gore and Clinton's insufferable fencing-in of Yellowstone (which has a massive supply of oil), so Seifert created legal boundaries around any use of wilderness and 'natural monuments', and stopped commercial use of unused wild land. Use of such land required official sanction and involvement (99).

In essence, fascism gives 'freedom' to men by making them wear chains, blindfolds, earplugs and mouth gags. Or, submit to death.

The aim of Nazi Green policies was "intellectual realignment". China has a similar form of 'realignment', as did Russia. That is, brainwashing. Note the requirements of modern environmentalists, who are clamouring for the same thing.

After Hess made his famous flight to Britain, the Nazi party clamped down on environmentalism and, for the last three years of Hitlerism, the green aspect of fascism was more or less inactive. "(The work of Greens) however, had long since left an indelible stain." (100) We could call it a 'dirty stain'.

> "Even the most laudable of causes can
> be perverted and instrumentalised in the
> service of criminal savagery." (100)

Yes, it is laudable to keep our lands clean and free of toxicity. Yes, it is commendable to keep waste to a minimum. These are solidly scientific and social aims when used for the good of mankind.

When those aims then start to become mythical, paganistic and without solid scientific reasoning, leading to the enslavement of the masses to a fairytale claim, it is time to clamp-down hard. (No, this is not fascism – it is self-protection and sheer common-sense).

People can be as insane as they want to be. They can believe any kind of nonsense they wish. But, when they insist we all comply with their insanity, we must resist and keep them at arm's length. The alternative is to make fascism a way of life, never knowing when the other aims of Hitler will come into force. (The current 'trans' delusion is just one example of fascism).

Alignment with the 'final solution' has echoed down through the decades since Hitler, a solution currently in vogue amongst hardened Greens, governments, and in UN policy. It is to lose all sense of what it is to conduct genuine research and to come to genuine scientific conclusions. It is to accept fascist-style imposition of laws, taxes and controls that have no basis in reality. And it is to follow Nazis along lines of eugenics and genocide.

What else do we call it when the West refuses to allow Third World nations access to industrialisation and means of earning money to live on, and gloating at starvation, but a form of genocide?

Environmentalism has its roots in both Marxism and fascism. Both are dead by now, long buried, and their bones crushed and stuffed into sewage. Yet, some, hungry for power, want to resurrect them though they are dead!

Today's environmentalists, from Al Gore to cynical scientists who pretend to find global warming and climate change caused by humans, to governments who are using it as an excuse to dominate the people, are just reincarnated Hitlers. Their wish to see millions die of starvation is a sure indication of their vile desires. Fascism is evil. So why are so many people supporting it through a scientific scam?

Remember this: Green policies in Nazi Germany led to the holocaust, because Greens thought Jews were 'weeds' who did not deserve to live. There is no difference between this inhumanity and the inhumanity of Greens today who smile at the thought of millions dying through a starvation they will not help eradicate. Do you want to be a part of this deception?

"To explain the destruction of the countryside and environmental damage, without questioning the German people's bond to nature, could only be done by not analysing environmental damage in a societal context, and by refusing to understand them as an expression of conflicting social interests. Had this been done, it would have led to criticism of National Socialism itself, since that was not immune to such forces.

One solution was to associate such environmental problems with the destructive influence of other races. National Socialism could then be seen to strive for the elimination of other races in order to allow the German people's innate understanding and feeling of nature, to assert itself, hence securing a

harmonic life close to nature for the
future." (101)

This is the true legacy of ecofascism in power:

"genocide developed into a necessity
under the cloak of environment
protection." (102)

Still think the fascism that initiated and runs
environmentalism will not become genocidal? It
already is! The UN itself is associated with siding
with genocide in various countries, especially the
eradication of Israel.

What would you do if Greens, backed by a
fraudulent UN IPCC and power-hungry
governments, told you that people in the Third
World are struggling to get a big share of your food,
and that this struggle will harm your children?
Would you agree to their death by starvation? Is
your answer 'No'? I sincerely hope so, because the
Germans fell for the same lie years ago!

Death by Starvation
Look again at what Greens are saying to
themselves! They have been talking for years about
allowing poor people to starve! If you support them,
you are also supporting starvation of parents just
like yourselves, who only wish the opportunity to
feed their children... the same children with large
dark eyes who are featured in those charity TV ads

and reduce you to tears. Cry for yourselves, friends, because you are targeted, too!

Will you now look at the screen, into their eyes, and say "I believe in fraudulent claims and bad science, so you have to die!"?

It is said that "those who want to reform society according to nature are neither left nor right but ecologically minded." (102a). That is true. And this is why I say that today's Greens are both Marxist and fascist.

They employ the ideas and means of the two greatest evils known to recent history. Sufficient quotes have been given to show the very real link between then and now.

Marxists have successfully poisoned people's minds against capitalism, when any problem with capitalism is not the system, but with some who are leading it. It is a simple matter of not putting up with flaws and making sure it works properly. But, Marxists want to eradicate the whole system (except when it gives leaders power and money). It is like killing a dog because it has one flea on its back!

The "deepening public concern" for the environment (120) is not concern for the environment at all. Most people do not have a clue about the environment, as we have already seen. What they are concerned about is their safety!

And if they must do what Greens tell them to do to remain safe, then they will support Greens, who are using a mythical doomsday scenario for precisely that purpose. It is the old group dynamics at work again.

Chapter 3

Marxism and Fascism Combined in Today's Green Movement

Stalin – the Green's Favourite Uncle

George Bernard Shaw, a socialist, and big fan of Hitler, once said: "The state has a right to kill you." Fascism has always been popular with the power-hungry. That is why it is dangerous. (103)

We have looked at the Marxist and fascist roots of Green politics (it is not really about the earth). In this chapter we will see that both movements are combined in the current environmentalist movement, and that leading greens, such as Al Gore, Tony Blair, Gordon Brown, the Clintons, Obama, the UN and EU, are Marxist and fascist all at once. This means Marx and Hitler are influencing modern day movements and all who believe their lies. (Few people actually believe – they are just sheep, made compliant by their ignorance).

The uninformed shout-down anyone who says Al Gore's infamous film is like neo-Nazi propaganda. They would certainly object if we say it is also tinged with Marxism. Look at how he operates, what he says, and his aims, and the reality of the combination is made apparent. (See my 2009 book).

This is not about Al Gore as a personality, conniving though he is, but about facts, and the way he is forcing everyone to obey the Green mantra which borders on insanity. I do not care what he is like as a person, so long as he does not force me to obey his myths!

The lies behind Gore's claims have not gone unnoticed. Sterling Burnett,* who comments on Gore's deceptions, is said to be a "man who works for an organisation which takes huge amounts of money from Exxon Mobile" (104).

His comments, and those of his critics, came in 2005. (*He is a Senior Fellow at the National Centre for Policy Analysis. From 1998 to 2005 it received just over one third of a million dollars from Exxon, peanuts compared to what Gore and Greens get from socialist governments... and even from oil companies).

This is an obvious and usual ploy used by those who tell lies – take attention from yourself by hitting out and dropping hints about something supposedly bad in a person's background.

Why not see who gives Al Gore his much greater funding? It would prove to be far more interesting! If the accusation is that someone who takes money from oil is bound to lie or be prejudiced, then what does that say about Gore, who also takes money from oil, but on a larger scale?

Very often, if an individual says something others do not like, they automatically call him 'fascist' or 'Nazi'. Most of the time the accusation is completely off the wall. The only way someone could legitimately be called a fascist is if his life, words and work mimic Hitler-style signs and symptoms. This we find in abundance in Al Gore (and others already mentioned).

Today, except for the currently missing <u>direct</u> genocide (indirect is already in motion), Gore's aims and techniques are those of Nazi Germany and Marxist Russia, all at once. Right now we are at the Weimar stage (just before the Second World War), But, the next stage is already planned; poorer nations are presently having their funerals finalised! There is no longer any doubt about that. He is literally getting away with it, because the naïve and the complicit follow him like lap dogs, rather than take a good hard look at what is going on. This makes them just as guilty.

Burnett compared Gore and his film to "the Nazi propagandist Joseph Goebbels". And he is scarily accurate. Argue as you wish, Gore is Marxist in ideology and fascist in his operating mode, and those who shouted Burnett down in a Fox News broadcast proved their ignorance, as well as their willingness to be vicious without cause. That is what ardent Gore-ites are like. They are willing to do anything to propagate Gore's lies, because they already have a socialist mindset that cares nothing for real proofs and research... mindlessness.

Notably, a spokesman for the World Wildlife Federation was a featured opponent... listen to 'Greens' at your peril! The presenter said we ought to rely on the facts, but, typically, presented none! She said she 'agreed totally' with pro-Gore fans, which shows she relied not on facts but on information given by Gore; fraudulent propaganda. In essence she ditched her journalistic skills in favour of something that has no truth in it. This journalistic attitude is now widespread and is called 'fake news', employed routinely in the standard media.

All who follow Gore show a refusal to look at actual facts; their belief is entirely based on propaganda and emotion. What this proves is not that people are stupid, but that well-orchestrated lies can be made to sound like the truth! And, as most people (including Gore) are not specialists in the sciences quoted, or have any idea about the genuine research process, there is no way they can know what the truth is.

What we are seeing is 'payback time' when the mass of Green followers grab any opportunity to push their own agendas and their own brand of politics. It has nothing to do with being 'Green'.

Burnett correctly told the audience that CO_2 is a natural and essential compound, not a pollutant. He pointed out that greenhouse gas theories are wrong because they assume the atmosphere is a closed system, which it is not (see relevant chapter in the 2009 book). If it was a closed system, logic and science would tell us temperature would rise exponentially. But, it does not - Greens deliberately do not take the many variables in the atmosphere into account. Facts don't matter!

The audience only quoted general statements already given to them by Gore! No hard facts, just passive quotes from the source (Gore) that gave them! If you go to the Helium website and scan the articles on global warming, almost without exception (the exception being mainly my own), you will find repeat after repeat of propaganda, not truth; emotion not hard fact.

The writers really mean what they say, but they have no idea they are just repeating false propaganda. They honestly think they are telling the truth. One of the major tasks of Marxism and fascism is to make people think they accept things by their own volition. Once they achieve this sleight-of-hand, people will do anything they are told, including the unthinkable. No? Then why are leading Greens clapping their hands at the deaths of millions in the Third World?

One critic of Burnett tried to poison the well by saying oil companies are involved with Burnett. So what? Al Gore has stocks in oil himself, giving him personal profits from oil (whereas Burnett does not get personal gain from oil), which makes Gore an hypocrite. The oil argument should be ignored; just look at the facts. Oil is not the bad guy here. And if you are playing the 'oil-peak' card, forget it. (I kick this myth out of the field in the 2009 book).

What we are seeing is the blatant telling of untruths and the spreading of those lies in the form of propaganda. When dissidents voice their view, even when giving actual facts, they are ignored or made to look stupid in public. That is how Hitler began his campaign of terror. His verbal arguments quickly gave way to engineered social disapproval and then violence and death.

Gore is on the same road. Perhaps he does not even know it yet, but he is. And the road follows the same trajectory as Hitler's, even down to spiritual/pagan exercises that are prevalent in the modern Green movement.

A book by John Foster brings this together, showing "Marx's neglected writings on agriculture, soil ecology, philosophical naturalism and evolutionary theory." (105)

Richard Levins of Harvard University said Foster did not write his book merely as a tribute to the past "but as an integral part of current issues. He demonstrates the centrality of ecology for a materialist conception of history, and of historical materialism for an ecological movement." (106)

This was echoed by Carolyn Merchant of the University of California, Berkeley, who links the past and present: "Should be of interest to all who care about the fate of our 'vulnerable planet'." (107) Therefore, the idea that Gore has nothing to do with Marxism or fascism is naïve.

The publisher's blurb also makes the connection between Marx and Greens: "This new account overturns conventional interpretations of Marx and, in the process, outlines a more rational approach to the current environmental crisis... (challenging) the spiritualism prevalent in the modern Green movement." (108) Note; There is no 'crisis' except for the fake crisis used to gain Green/socialist power!

The author himself, in a recent article (109), covers much ground I have already covered in chapter one of this book. He says "Ecology is often seen as a recent invention", and goes on to point out that it has been "already expressed in the work of Karl Marx and Frederick Engels." His statements on metabolic rift are also covered in chapter one of this book.

"Today," says Foster, "the ecological issues that Marx and Engels addressed... read like a litany of many of our most pressing environmental problems..." (The list of these so-called 'problems' is already given in chapter one).

He brings in the source of modern claims that human beings are responsible for introducing massive damaging amounts of CO_2 into the atmosphere, even though what humans produce is actually in minute proportions. "No-one at that time (referring to Tyndall's work)... suspected that the greenhouse effect, interacting with carbon dioxide from the human burning of fossil fuels, might lead to human-generated global climate change – a hypothesis not introduced until 1896 by the

Swedish   scientist,   Svante   Arrhenius."

That such a broad statement is scientific nonsense does not bother Foster, because he reads past ecological statements as absolute truth, though modern science disproves them! Nevertheless, he correctly adduces Marx to be the precursor and driver of modern green policies:

> "Today, the dialectical understanding with regard to nature-society interactions that Marx and Engels embraced, is increasingly forced on us all, as a result of an accelerating global ecological crisis, symbolised above all by global warming."

Thus, as I keep saying, Marxism is 'forced on us all', because of a 'crisis' that does not exist, but is a projection into the future of Marxist ideas on ecology! The whole Green movement is screaming forward not because of current problems, but because of perceived future problems dependent on socialism being true!!

Why accept Gore's bizarre and untrue claims without proof? If you accept without questioning, it shows your aims are political and not 'Green'. It also shows a willingness to become fascist by degrees.

Marxists can be Marxists and fascists can wear all shades of brown, but once they try to force me to comply with their zany, nasty, controlling ideas, I will resist! Why, then, do so many ordinary people follow like lambs being led to the slaughter?

This is like the whole world shouting insanely about a Martian invasion in five years' time that will bring doom to the human race if allowed to happen... even though there is no evidence at all of Martian life, let alone an invasion! (With one exception – the film, 'War of the Worlds'. But, common viruses took care of that lot, so do not worry!).

Only the insane would gather all world governments to 'combat' this mythical invasion, right? And only fools would allow it to go ahead anyway, committing zillions of dollars tax money (OUR money!) to the cause. Yes?

No. Politicians who do this are cool and calculated, because they know extra taxes will give them more power and more wealth. They are not fools, just modern politicians; few of whom are genuine or honest. They don't care if what they claim is found out to be untrue in umpteen years' time, because by then they will have gleaned their power, status and money!

Marxists believe that what Marx said a long time ago can never change and his beliefs must hold true for all time, even though Marxist countries have shown their incompetence and inability to act out in the real world.

Thus, Foster actually believes that the Jevons Paradox, devised when the industrial revolution came about, must apply throughout the ages. The Paradox says that when an industry increases its production, even if updated techniques are used, it also increases the problems, by using more natural resources and energy, and putting "more strains on the biosphere".

It is like listening to a man suddenly thrust forward from Mid-Victorian England with no current knowledge of advances in technology. Because that is exactly what the ideas are – out of date and irrelevant! As you will see in my assessment of 'Transition Towns', the claims to 'oil peak' are just as outdated and irrational. In a very small way I placed a container of creosote into a supermarket plastic bag and left it outside. It is the kind of bag we have all been stopped from using (or charged money for), because they take "hundreds if not thousands of years to disintegrate". Well it took just two years! The bag disintegrated to nothing. Also, many similar bags are now made of materials that disintegrate quicker. But, the facts don't matter! Forcing the public to obey and charging for use is all that matters!

Like anyone else caught in the Marxist mindless thought-trap, Foster, referring to Brett Clark and Richard York, says "Technological development cannot assist in mending the carbon rift until it is freed from the dictates of capital relations." (110)

The 'carbon rift' does not even exist. Therefore, the argument from Marx does not have validity. And the phrase "freed from capital relations" is the talk of a capitalist-fed Marxist living in the free West, who has never lived under an oppressive Marxist yoke! The idle-rich Marxists live a great life, but deny it to others.

It is significant that leading Western Marxist thinkers have expensive houses, expensive salaries, and the usual trappings made available through being capitalist in life but Marxist in thought! All made possible by the sometimes mistaken tolerance of Western society towards people who wish to mess with society's well-being.

The idea that Marx has nothing to do with modern environmentalism or with people like Bill Clinton (Director-General-in-waiting-he-hopes of the UN and its fraudulent IPCC), Tony Blair (President-in-Waiting-he-hoped of the secretive and dictatorial EU, but never got the job) and Al Gore (King-in-waiting of the whole world), all of whom are Marxists, is laughable, and indicates a deep degree of ignorance on the part of people who support them and their causes. (See Appendix, 2009 book on the UN).

But remember, these people, though Marxist, are also fascist (both are socialist). They push through their beliefs, using laws and their unfortunately high offices, like nailed clubs, to force people into submission.

Rift Between Men and Their Brains

"Recent research in environmental sociology has applied Marx's theory of metabolic rift to contemporary ecological problems such as the fertilizer treadmill, the dying oceans, and climate change. Writing on the social causes of the contemporary 'carbon rift', stemming from the rapid burning up of fossil fuels, Brett Clark and Richard York have demonstrated that there is no magic cure for this problem, outside of changes in fundamental social relations."

Read things like that fast, without genuine research, and you will get caught up in the same manic rush to return to Medieval England, forgetting just how comfortable you are in modern society! Look closer and you will see the 'rift' is not between men and nature, but between men and their brains.

Though there may be a problem with using vast quantities of chemicals, there is no 'contemporary problem' with fertilisers that technology cannot amend. The move to replace harmful chemicals with non-harmful substances is already being addressed.

What 'dying oceans'? There is no evidence for such a claim. And 'climate change' is just normal and cyclical! There is no problem! A zealot of climate change shouted out that the current 'global temperature change' of less than one percent must be stopped now, or we will all suffer!

For goodness sake, as one scientist says, who can register such a minute temperature change when they go outside in the sun? And, the figure is so small, a mere fraction of a whole point, as to be a very silly banner under which to call for worldwide panic! But, who cares? Do YOU just accept the absurd panic? Or, do you think independently and rationally?

A less than one percent change is of no significance, especially as it is not even global. All the evidence suggests the exact opposite of damage: global increase in health, wealth and food supplies!

But, wait, this will not suit the rampant Marxist who wants us all to live miserably under his cold, grey regime! Forget the benefits of temperature rises, just concentrate on the suffering that Marxists wish to occur, so that they can be hailed as saviours of the world with their outdated and failed political system!

Wait again! All this applies only if there is such a problem to combat! Climate change happens! It can change from town to town, state to state and country to country. I live in a valley about five miles from an isthmus with fairly flat terrain. I know that when it is raining and cold in my area, just five miles away the isthmus is often warm and sunny!

There is no 'global warming' and no current way to prove it exists. But, Marxists do not care, so long as they get their huge grinding machine to work, crushing the will to live. Go back in time if you can and ask the jolly Soviet people under jolly Stalin if they would like to see Marxism survive!

You might be surprised to find they were not so jolly after all. Maybe Western academics do not quite understand the difference between Marxist thoughts in the head and living under real Marxist regimes.

Stop the Lorries!
Amusingly, you can see the absurdity of applying the Jevons Paradox to the modern day, in a TV advertisement for washing powder. A child narrates it, and says that if the powder is used, it will halve the amount used, halve the water used, and will therefore halve the number of lorries used to carry the powder to supermarkets and the amount of CO_2 emissions. Great stuff for the Greens. But, what garbage!

But, look at it again, because it encapsulates the Marxist theory of resource replenishment. All that will happen in reality is that the manufacturers will increase production and find other ways to bring in the shekels. And the number of lorry journeys will not halve, but remain the same, or grow into more.

Do you think lorry drivers' unions will accept a sudden halving of their contracted hours? Or, that employers will accept their fleets will remain idle when they are paying for them? So, amount of powder production increases, the lorries run as usual, and CO2 emissions remain the same. Oh dear! Nice advertisement though.

As others have already observed, Marxist green ideas now being put forward might work well in small pockets of fairyland, but they do not translate into large-scale working scenarios. The nice lady with the wool hat and gumboots next door can work it well and live off her own produce, but it does not work in whole countries. So, why should they work - God forbid - on a global scale? Yet, Marxism is the model for Gore, Blair, the Clintons, their sycophants and ignorant followers.

Green = Marxist/fascist power. "The goals of human freedom and ecological sustainability are... inseparable, and necessitate for their advancement, the building of a socialism for the 21st century." (113)

You can see that the aim is not green at all, but Marxist. To put the words 'socialism' and 'human freedom' in the same sentence is a logical absurdity and an historical lie. Look at any truly Marxist society, past and present, and you will not find human freedom! You will find regression, repression and oppression, poverty, sickness, overvalued 'results' and impossible goals, held together by violence.

Again I urge you to read the Gulag series by Solzhenitsyn. The huge tri-volumes are heavy reading, but in them you will find the reality of Marxist society, in great detail. There are no gains or benefits, except for leaders and sycophants.

For the rest, the 99.9%, life is just a grind, a daily torture; aims are mythical because they are unattainable; results are faked to fit the mythical aims!

Read the almost comic way forestry workers and bosses fiddled the figures to make them look like they adhered to the massive ten year plans. They did it by pretending to send felled trees to other parts of Russia. Then, those they supposedly sent them to colluded, and 'sent' them on somewhere else. Nothing worked but the delusion was maintained because people were afraid to tell the truth!

In the end no-one knew where the pretend trees were... but the 'results' proved the 'aims' to be correct. And, the ineptitude of Marxist-Leninist ideals were almost the ruin of Russia. This is the ineptitude we will inherit if Greens get their way. Go on, read the Gulags and see what Marxism really does to a people! Why resurrect a dead body when the flesh has already rotted off the bones? Why eat Green roots when they have rotted? We need Marxism like gas-chamber Jews needed Hitler. And I do not say that lightly, or with any sense of humour.

Yes, Marxism is Dead... But Hey, Let's Give it Another Shot!

Everyone applauded when the Berlin Wall came down. Just a few years later those same confused people are rebuilding it again, through Al Gore et al!

> "Despite its academic stature and its pervasive impact on all species of radical thought, Marxism as an explicit political philosophy finds little favour in most eco-radical communities. If the spectacular ecological and social failures visible in all communist and ex-communist countries is not dissuasion enough, many environmentalists are ready to dismiss Marxism as irredeemably sullied by its humanistic heritage. Yet, while denouncing Marxian doctrines, radical greens retain, and indeed often extend, the Marxian critique of capitalism."(113)

So, Marxism, like materialism/atheism, is dead, but Marxists want to dig up the body and pretend it is still alive. Whilst some radical Greens say communism is dead, they nevertheless follow the Marxists and use the same dead body in their regimes, though all known Marxist regimes have failed and caused human harm. Can you see a flaw in their logic? Even the author quoted above, a committed environmentalist, says in the blurb for his book:

> "Those who propose a radical environmentalism (e.g. Marxist) unwittingly espouse an ill-conceived doctrine that has devastating implications for the global ecosystem... an ecoextremism... that would, if enacted, result in unequivocal disaster."

This is exactly what the Green movement proposes!

The academic quibbles are interesting but futile, because it is ignorant people like Gore et al who lead the way! They are ignorant because they have themselves believed the ideas of extreme Marxist and Green eccentrics, whose common sense just dribbled out of the broken bottle! They are ignorant because though real science debunks what they have been led to believe, they refuse to even look at the counter-evidence, let alone amend their views.

Of course, if they did listen to truth, then their amended views would leave them without a cause, for there is no crisis and no problem! With it would go their worldwide status and influence and income. And with all that go the livelihoods of the pathetic scientists who insist on foisting their bad science on the world just for the sake of funding their departments. Now that is the real problem!

In their fight against reality, ecoextremists have turned sense on its head by calling the Soviet Union the "ultimate form of monopoly capitalism" (114). It does not matter! Marxism is Marxism, and Marxism obliterates the individual and individual wellness, ruins whole countries, and elevates mindless fools to office.

I visited a tiny picturesque village in Istria, Croatia, a country only a few years out from its communist past. It is now trying to build an economy from tourism after enduring decades of communist dictatorship along with its intense poverty. For years, its citizens emigrated from Croatia to western countries, simply in order to survive. Today, all of its agriculture, including its long tradition of winegrowing, is natural (nowadays called 'organic'). Everything grown and eaten/drunk, is organic, as it always has been.

Now that Istria is at last making it as a tourist destination, it is intent on retaining its organicism and culture, through touristic 'capitalists'. They see it as the best and most suitable way to keep the Istrian character and 'wellness' whilst building a sound economic base. That is one in the eye for Stalin! And certainly irritating to Greens, who want to destroy capitalism.

Yet, in that tiny village, with its crumbling houses, the name of Tito was painted along one wall a long while ago, praising his dictatorship! This was not the work of an ordinary peasant, but that of a minor official in the village whose bread was augmented by an even higher official. The village just kept its head down and outwardly agreed with the minor official. Because, that is how they were allowed to keep alive.

But, to see modern men, who have witnessed the horrors of actual Marxism in other countries, wearing the cloak of a minor official just to get their own way, is abominable. Such idiotic adherence to what is dead (and should stay dead) makes no sense, except in terms of personal aggrandisement and power.

> "While the modern green-reds promise to include socialist ecological failures within their analytic purview, 'capitalism's global destruction of nature' remains their over-riding concern." (115)

> "The essential aim of eco-Marxism is to use the environmental crisis to revitalise

Marxism. In practice, this involves two procedures: first, showing that whereas communism may be accidentally destructive, capitalism necessarily destroys the earth; and second, rehabilitating Marx and Engels as ecological thinkers."(116)

"Imaginative thinkers are even able to blame ecological disasters in the former Soviet Union on the capitalist world system." (116)

"The flexibility of contemporary Marxism is such that green-red thinkers can still set their sights on capturing the leadership of the entire radical environmental community." (116)

The idea that capitalism necessarily ruins the environment runs throughout Al Gore's and green arguments. It is like a mantra they feel they have to chant, whether or not it is true. True or not, it is typically Marxist. Note that Marxists want to dominate the whole Green scene, using fascist clubs to beat out the brains of dissidents.

Real Concern; Fake Reasons
Marxism is good at creating a state of frenzy in people, as a means to an end. This is what has happened with modern environmentalism. Until very recently, to be Green was to be a rather odd

but quaint person who liked to tiptoe around the countryside looking at fields and flowers, and maybe making sure insects in a pasture or two are protected from nasty road-makers. It was good-humoured and they could still use reasonable language and arguments. But, things have changed, drastically.

Now, Greens are not eccentric likeable people. They are nasty, strident, and highly political. This is most odd, because the majority of people who recently signed up to the Green cause, do so because of global warming, climate change and CO_2 emissions. That is basically it.

But, they don't join up or sympathise because they have a reason to do so. They do it because they have been scared witless. Marxists have done this, because scared people do not think rationally and tend to run around like headless chickens.

Sorry to say this to people who were once ordinary and nice, but it is true. They have become puppets in a red plot. They support Al Gore because they are frightened, not because they have truth on their side. They have been told, repeatedly, that they are in a pit soon to be filled in with concrete. Then they will die. In reality, they are not in a pit, there is no concrete, and they will not die!

They are victims of marvellous devious marketing. Though there is no truth in the Green agenda, it is made to appear true by manipulation of the media and academia. And now that governments have been brought into it, the Green claims must surely be true! Why? Governments are manifestly continual liars!

Marxists will stir the pot of fear until people listen to them and obey. And obey they will, just as Germans obeyed Hitler and Russians obeyed Stalin! For them, as now, it was all about survival. If it takes going Green to survive a coming global holocaust, then so be it!

But, nobody who follows this idea knows anything about what is supposed to be coming. Yes, they have the propaganda, but no genuine science. Fear and ignorance drives them to obey.

Did you know that when Stalin held his packed (by command) Party conferences, everyone was expected to clap for hours? The first ones to stop were reported to officials and suffered the consequences. Did those who clapped believe what they were told? Of course not. They were just too scared to say anything or to stop clapping a vile mass murderer.

Ask anyone who supports Greens to explain what the 'problems' are, according to Greens. They will repeat the mantra and the false facts, but cannot explain what they mean! Know why? Because they are not scientists and they rely on what Greens (most of whom are also not scientists) tell them. They don't check the stories.

There is no way they can distinguish scientific truth from lies. But, they support Greens anyway, because they do not wish to find themselves under twenty feet of water, or lost in clouds of CO2. The fact that both claims are false and are unscientific makes no difference, because fear drives them onwards.

They are, then, the Stalinist peasants of our modern day. I say this not in disrespect, but in sad horror. No doubt many who follow Greens do so because of peer pressure and the fear of not being one of the 'in crowd'. (For proof of this, see a funny but sad reference to an experiment by the magicians, Penn & Teller, in the 2009 book). Really, though, there is no excuse.

A TV presenter read a book on fascism by Jonah Goldberg, and was shocked by the implications (117). He said he refused to accept that what environmentalist Americans were doing was 'fascist'. But, as we all know, and as I have already said, if it looks like a duck, quacks like a duck and waddles like a duck... it's a duck!

The presenter did not want to face Goldberg's conclusions, but, commenting on how Greens work, he said 'Green is good for you, so we'll force you to accept it'. And it was Gore who started a rumpus by calling dissidents 'fascists'! All we are doing is responding, and reversing his accusation – but with truth.

Michael Mann (inventor of the infamous 'hockey-stick' graph used by Gore) reviewed the book (118) and also refused to face the facts, saying that if anyone wishes to bring down Al Gore's work, they should call it 'socialism' rather than 'fascism'. As if that would make any difference! What does he prefer to call the environmentalist wish to 'depopulate' by starvation? (Mann also tried to attack me because of <u>my</u> book, which pointed out his scientific 'hockey stick' nonsense!)

Though hating the label, he refers to the author: "He is correct that many fascists, including Mussolini, started as socialists" and that Goldberg says there is a "family resemblance": "It's family because American liberals are descendants of the early 20th century Progressives, who in turn shared intellectual roots with fascists."

So where is the argument? Remember the duck! Mann, of course, thinks Goldberg is off the mark, even though a review by Publishers Weekly says the book is "well researched." (119) (Which Mann's hypothesis was not!)

Goldberg rightly links Hillary Clinton with 'liberal-fascism'. The same label is applied to Democrats and the environmentalist movement. Again, rightly so, because that duck waddles through them all. The review says what others have said – liberals will stop dead just at the cover, because they will hate the link between fascism and the 'New Deal'. By 2018 Democrats certainly proved their fascist/Marxist roots.

But, look at what it all means! Look at what is said! Look at how it is being brought about! It is out-and-out fascism. Many liberals have called the book 'stupid', but then, they would! Throw as many spanners into the works as you wish – the duck is still there, plain as day, and it can be seen many times over in this book, too!

Like, or do not like, the word 'fascist' applies to Al Gore, it has no relevance to what he is really about. Test him. Check his words. Watch his actions. Notice what he projects into the future. Check out the science. Investigate his pals. See how he intends getting obscenely wealthier by scaring the life out of people.
And if you then do not see blatant fascism oozing from his every pore, well, you should find out if you need truth-spectacles. As for Marxism, this is obvious, throughout Democratic treasonous politics and how Gore thinks.

The aim of Marxism is to divide and conquer. Gore does this supremely. Fear is an excellent way to cause this division in people. Once they are made fearful they will want to find someone who can provide all the answers.

Oh, who is that we see in the distance? Is it a friendly Marxist? Yes! Is it Al Gore? Yes! Is he saying he has all the answers? Yes! Then we had better follow him and obey everything he says! Before we all drown in the non-rising seas.

Others say he is lying and the facts are not facts at all, but are complete lies. Should we perhaps check things? Oh, who cares. Shut up and get out!

So you see, Green supporters who are not scientists (the majority) have no idea that what they are being told is fake. Nor do they have the qualifications to check it out anyway. So, how can they do so?

Well, one thing for them to remember is that where there is more than one theory about any scientific issue, there is, by definition, no settled law of science, and no reason to bring in draconian laws and demands. As for 'consensus', in science this is not possible and opposes the true scientific spirit.

Therefore, hang back and wait to see what happens. The last thing you should do is rush forward screaming about doom. By doing that you make Al Gore's job of scamming you easier and you will lose your ability to think clearly in the process. His job is to make lots of cash... off you.

The risk capital company he has joined wants him to move fast, because that is how risk capital folks make their biggest profits (see 2009 book). By running fast, then, you will be putting billions of dollars into their personal bank accounts, making Al Gore a very happy bunny indeed.

And by the time you discover it was all a lie, Al Gore will be on his own large island in a very luxurious mansion, where there has been no sea level rise, sipping a cocktail, continuing to reap benefits from his stocks in oil, etc., and using a lot of electricity.

And guess why he will be laughing? Because the guru of Greens is a man named James Lovelock. He is the one who made Gaia popular and said we will die soon. Yet, he also said it is already too late, so there is nothing we can do to stop the coming (mythical) global disaster. Why, then, pay extra taxes to stop something we cannot stop? Why allow the UN, EU, and governments to restrict our choice and use of energy? Why let them cut off our freedom of speech and choice? Why destroy industries and economies? Why indeed!

Run for the Hills!
Can you see what I am getting at? There is no problem and no coming global crisis. And even if there were, it is already too late according to Lovelock! As we have seen, there is only one answer – Marxist rebirth with brown-shirt midwives.

The objective is domination of people on a scale Hitler and Stalin could only dream of. And it is being done through the Green movement. Unless everyone stops being minions to the super-marketing men, they will suffer. Marxism causes suffering. Look at Marxist states today and what do you see?

Do you really want to join them? Well, you will, very soon, if you support the current environmentalist cause. And do you just hate those leaders who think they can bully you into submission?

Why, then, do you submit to the intellectual and legal bullying of Al Gore and other fascists? By repeating their mantras and waving their banners, you are already in submission and oppressed.

The 'But' Buts

"But, global warming is obvious!"
Oh, is it? If a grid is placed over a map of the world, each 50 miles square, and temperature recorded at exactly the same time in the centre of each square, you would get a different temperature for each one. Do this every day for, say, ten years. What you will find would prove beyond doubt that your statement is nonsense. And how do you know anyway? Have you plotted all the temperatures in all the world on the same day for a reasonable period? (Ten years is not reasonable – you need centuries, not decades). And, have you taken into account the

many variables that would change those individual temperatures? What about extremes? What about temperature boxes placed near central heating units outside houses and factories, skewing measurements?

"But the sea levels are rising!"
In some places there may be a rise of millimetres, which is normal, but there is no dramatic rise sufficient to fuel panic! In some places the land is rising! In some the land is sinking. Both are slow processes. But, sea levels are roughly the same as they always have been, or, are changing in an expected way. A British law court found that out and said the claimed dramatic sea level rises are a lie. Maybe, though, you prefer to panic. Oh well, that is up to you. (I also checked the story put out by Gore, that certain South Sea islands were already sinking under high seas, so the entire population left for New Zealand. The New Zealand government told me it was a lie – there were no sinking islands and no influx of islanders!)

"But human beings are producing too much CO2 emissions!"
If you believe this, without checking the facts, I must ask "Do you really know what you are talking about?" No, I do not think so. There is no proof whatever that CO2, or its levels, are a cause for concern. And there is no evidence that humans are creating some kind of evil poison to kill us all! It is

all a myth designed to make you comply with Al Gore and his Mean Green Machine.

Al Gore does nothing at all to stop his own CO2 emissions, but he demands that everyone else does! Thousands of scientists oppose the propaganda about CO2 emissions. Far more than the '2,500 climate scientists' who gave reports that made up the 2007 IPCC Report (the one you are now in a panic about which was written mainly by non-scientists). An almost identical Report was presented by Obama, in June 2009... same lies and fraud!

Did you know that the '2,500 climate scientists' were not all scientists... indeed less than 45? And did you know there are far less than 2,500 actual climate scientists in the whole world? No, I expect you do not. This is because you are being duped by Al Gore and the UN. The UN is Marxist (see 2009 Appendix) through and through, and that is why they are using deception to bring you alongside. Once you are on board, the whole scene will change and you will know the truth about their real aims. Then, it will be too late.

Green Marxism

> "Green Marxism? The notion will seem
> far-fetched to many. Yet, capital is the
> enemy of nature, and Marxism is the
> discourse of anti-capitalism. Any real

confrontation with the ecological crisis
will require the Greening of
Marxism."(121)

The authors of this book have missed the boat!
Marx was already 'Green'. It is just that Stalin was
too interested in killing a few million people to
bother much with Marx's ecology. What we see
today is Marxism reviving and putting into place
what Marx had already written about (from the
safety and relative comfort of his home in a free
capitalist London!).

However, the Green movement is no longer just
Marxist. It is also fascist, and this is why another
reviewer of 'The Greening of Marxism' was able to
say "a green future lies neither to the left nor to the
right..." (123) What is neither left nor right?
Marxism with fascism – left and right combined!

Greens in Politics

"many of the early leaders of Greens
were former socialists and left activists..."

"The Green's initial membership base
was... pushing to the left."

"Politically, the emergence of the Greens
was a big advance over the
environmentalism of the 1970's... The
formation of Green parties created more

space for the left. Their electoral
challenge created a real potential for
opening up a complete restructuring of
the political scene which could end the
dominance of the ruling-class parties... a
left dynamic developed around the
Greens... the left was strong in German
Greens from the start."(123)

It has been said the only movement that can
transform society today (whether or not it wishes to
be 'transformed') is "political ecology". (124) It was
part of Obamaism.

"More and more militants and
intellectuals, coming out of the (Marxist-
inspired) labour movement, are finding
one another in the ecological camp...
they found in the political ecology
movements a 'family resemblance' (to
classic Marxism)... (including)
materialism (and) dialectical historicism...
The greens share with Marxists the
conviction that they have arrived at the
moment... when a particular order of
things brings us so close to catastrophe
that a great transformation is required."
(125)

Once again, we see that a 'transformation is
required'... but by whom? Not by the people, but by
those whose minds are locked into the mindset

forced upon them by their political theories...
Marxists. 'Change' was the big word used by
Obama.

Unable to see outside their red box, they force
others to join them, using deceptive propaganda,
not truth or genuine science. And ordinary folks will
know no different.

Wait, Fascism is there, too!

> "Fascism is difficult to recognise because
> it is so poorly understood and because its
> nature is masked behind collective
> denial."(126)

Hitler is not the only fascist! Indeed, many modern
fascists show a very genial face to the public (which
is why Goldberg's book has a jolly round smiling
face with a Hitler moustache). For example, take
the Clintons (yes, please do!), and Al Gore, or Tony
Blair. Who would suspect they were fascists? Well,
they are both Marxist and fascist, so if nothing else,
they are up to date.

Remember that quote above, in which both awful
movements are combined into a new kind of middle
ground? That is the 'synthesis' of the thesis and
antithesis, Marxism and fascism. The skinheads,
Nazi troops, and communal local czars, have been
replaced by men with big smiles, fast talk and well-
tailored suits. But, their thinking, words and actions

are all clearly Marxist-fascist. Amusingly, both sides condemn Nationalist parties as a smokescreen!

As I have already said in chapter one, you can get rid of the outward trappings of an ideology, but not the ideology itself. Especially if that ideology gives a person something he cannot otherwise have, such as power, money or status (even those local bossy people in local councils are sometimes called 'little Hitlers'; they get their kicks from having an illusion of power). In other words, in such a case, the ideology is a form of robbery and abuse, using deceit as a weapon.

> "The defeat of Hitler and the Axis powers in World War Two, meant the military defeat of fascism, but an ideology cannot be defeated by military power alone. Ideas linger. They are reborn when the time is right again, or they can come out of hiding in strange new shapes. The major scandal of contemporary thought is that, despite World War and the Holocaust, the intellectual heritage of fascism has never been repudiated."(127)

It is said that the "current culture" and the "implosion of communism... offer a fertile breeding ground for the revival of fascism." (128) As we have seen, though communism as a whole is dead, Marxism is now rising out of the ashes in new ways, as is fascism. Maybe it should have a cute name, like 'MarxiFascism'! This new-oldness is easily detected in current USA politics and academic circles.

Note that the "current culture" allows the vile movement to breed again. It is particularly found in the PC movement, a virulent virus that destroys nations and the ability to think freely. The PC way of thinking is now general bad practice in all our universities, where both leftist and rightist professors rule what people think and do, dismissing staff who will not bow to their demands and perverted academic statements, and get rid of students who cannot tolerate their oppressive regimes. Proof of this perversity is the way all students who enter university must now prove their liberal orthodoxy by attending 'diversity' lectures. Thus, they begin their life of socialism and are forced to retain it, or be marked down.

Deviance from this enforced norm results in castigation and even removal from the university. This is what fascism does. It cannot win by talking truthfully and gaining adherents by reason of good argument. So, they just use a club and beat everyone on the head until they submit.

"Calling someone a fascist is a common way of insulting an opponent in political debate" ... or, no debate at all. Several movements use this tactic, whether terroristic, sexual or political.

So, is my book merely insulting others for the sake of covering up my own fascism? It is a fair question, but not based on fact. My view is that anyone can say anything, so long as I am given the same courtesy. The fascist view is simply to silence people, call them names and ban their views. See the difference?

The Dark Side of Green
Many of the tenets of fascism are found in the environmentalist movement. Much of it is not meant to occur, but the panic spread by Al Gore pushes Greens hard to be fanatics, keeping up the pressure to "do something before it is too late". With no time to think properly, they become agitated and strident... and the marks of fascism start to be seen. I am not insulting people, but warning them of something they may not be aware of. They are certainly ignorant of the facts.

> "Darwinism gave new notions of heredity, race and environment. The Nazis took Darwinism to its natural conclusion: If you can breed better sheep by selective breeding, why not human beings (eugenics)?"(129)

Note again the involvement of Darwinism/materialism/atheism. Darwin never designed his theories to impinge on social theory, but that is what happened anyway (because the original hypothesis contains the seeds to do so), sparking off many new movements, including Freudian psychiatry (now debunked), Marxism and fascism. They saw in Darwin's work something to give their ailing ideas a fresh boost, the impetus to soar high. However, this is not about Darwinism but about how budding Marxists and fascists used his ideas for their own ends.

The fascist notion that Christianity is null and void comes from "Nietzsche's criticism of Christ's teaching of loving the downtrodden." (129).

The environmental movement (that is, the fake kind) is intolerant of anything not aligned to their theories, including Christianity. Nasty things must be done if we are to 'save the planet', and that includes the opposite of Christian tolerance and love – the killing of millions who are deemed to be 'worthless' or 'weeds'. Or, as some environmentalists call them, "a virus".

These nasty things must be done to keep on track, to be true to the cause. Sounds like fascism? That is because it is fascism.

Fascism is a worldview that takes in all of life (and is, in essence, a religion, as is atheism). Many suppose it is the 'polar opposite of the left wing' (Marxism), but "while there are differences there are also similarities"

That is why Edgar J Hoover (despised by the left-winger sycophants of Obama and his fake government, because he highlights their own fascism) once referred to communism as "the Red Fascism", because it mixes both vile movements into one.

Though fascists want to build something new, they also cling to ancient paganism. Note that modern environmentalists openly say they will build new societies, getting rid of old ways of living, old traditions and transcendent religions. Paganism already underpins their movement. They do not ask us if we wish to be dragged along with them – they expect us to just submit! Hear the shrieks coming from Green fanatics because President Trump refuses to join them; they unashamedly display all the marks of fascist-Marxism.

Fascism normally wants competition nationally, but Marxism sees its 'struggle' as international, making everyone the same.

The similarity is that both use force, propaganda, rejection of free-thinking and the imposition of oppression. These are very evident in environmentalism.

Four Major Tenets of Fascism
1.   The idea of a personal God is hated. Why? Because the Christian God is separate from His creation, not a part of it. Thus, fascists hate Biblical

doctrine but accept existentialist people who call themselves 'Christian', because they are in the 'here and now' on this earth. This ties in with their view that everything should be organic, with a mythological unity of nature.

During the Nazi era, these existential 'pretend' Christians were Hitler's allies, because they wanted to be rid of a Hebrew-based set of beliefs. So, they agreed with the death camps and their 'solution'. They were, then, fascists themselves. This same hatred for authentic Christianity is found in Al Gore's worldview, because he blames Christianity for the 'ruin' of 'mother earth'.

2.   To get rid of the idea of a transcendent God, the Nazis reverted back to paganism, because it both destroyed Christianity and invoked mythical gods that were of the earth. They chose Wotun to lead this paganism, because he was of Germanic origin. Today, because fascism blends with Marxism, the pagan deity is Gaia, Mother Earth, because 'she' is international. The modern environmentalist movement follows, and paganism can therefore be found mixed with Green policies, even amongst otherwise highly-placed academics!

3.   Fascists, like Marxists, hate personal worth and individuality (except the leaders, who think very highly of themselves!). There must only be a national or international identity (though the former position is now being removed in favour of

globalism)! The culture must take priority, and individual worth is defined as arising out of the culture. There can be no opposing views! Note that Al Gore either ignores opposing (genuine) scientific arguments and opposing scientists, or refers to them as 'deniers', who are outside the cultural norm. So, he is using fascist techniques to silence people.

"This glorification of culture and ethnicity of course led to virulent racism the Nazis have been noted for." (Fascism, p 4). Think there is no racism in environmentalism? Look at what Greens wish to happen to the Third World poor, which is also stated by famous people such as Prince Philip! Is death by gas chambers any different from death by withholding means of growing food, or allowing starving people to die? No. Well, that is Green philosophy.

4.   Fascists see a human unity with nature, similar to that found in "primitive and animistic cultures". (Fascism, p5). This is paganism, not science. Man is one with nature. "What is generally not known is the fact that this doctrine led the fascists to hold a very zealous view of environmentalism. They enacted significant environmental protection programs... The state not only was to bring about this unity (hence the tolerance for totalitarianism) but the state itself was conceived as the embodiment of culture. The state was to be a living organism in which each individual

would find fulfilment and purpose, like cells in the greater body." (Fascism, p 5) This is exactly what we find within environmentalism, the brand that is working worldwide, intolerant of any other belief or genuine scientific facts.

I have listed four distinctives found in fascism, because "one is not a fascist unless he adheres to the entire corpus of ideas" (those listed). (130) Every distinctive of this corpus is found in modern environmentalism and in Al Gore's speeches.

> "In the 1930's avante-garde artists
> shocked the bourgeoisie with their
> aesthetic theories that glorified violence
> and the release of primitive emotions.
> Today, if you like examples of early
> fascist aesthetics, simply go to the latest
> Hollywood blockbuster, turn on MTV, or
> go to a heavy metal concert. Here you
> will see realized the fascists' artistic
> ideals: pleasure from violence; the thrill of
> moral rebellion; the cult of the Aryan
> body... the masses of teenagers slam-
> dancing as Metallica sings 'Scream as
> I'm killing you'! Such art is the
> quintessence of the fascist aesthetic."
> (131)

Those caught up in these 'aesthetics' are the people who support the crushing ideals of modern environmentalism. So it should be no surprise that their idea of 'culture' should be that which is created by fascism.

> "Contemporary mass politics is very different from the democratic ideals of Madison and Jefferson. Instead of rational analysis of issues and reasoned debate, our political discourse turns on image manipulation through the mass media... Visual images take the place of language; emotionalism takes the place of logic. Politics is trivialised; citizens are manipulated, but they are molded into a common will. This was Goebbels' dream." (132)

Look around you. Is this not what you see? Al Gore has been invited to debate his corner but he refuses to do so. Instead, he and Greens push fake shots of polar bears and penguins, and bits of ice falling into the sea. One of his greatest critics is not a scientist who has no Green background, but one who was a leading Green and scientist, Bjørn Lomborg!

The US government, until recently sadly led by Marxist-fascist Democrats, refused to consider alternatives to drastic Green action (on the advice of Gore). They asked for proof that the IPCC report was true, and though it has was shown to be false, nothing has been done. But, because President

Trump rightly rejects their insane Greenism, they are furious to the point of stupidity and violence.

The initial step should now be, if governments were being rational, to immediately stop all efforts to extort taxation on the strength of 'climate change', and to stop all actions that come from belief in global warming. But, rationality is not happening. This is because they are fascists working on the basis of 'Goebbels' dream'.

> "Moral issues are today almost
> impossible to discuss in objective
> terms... the majority of people today
> have no concept of an objective morality
> that transcends the individual and
> culture, Morality is reduced to social
> utility or the assertion of the will. This was
> precisely the Nazi ethic." (133)

What we have is a reliance on relativism, even though it is a forlorn philosophy bringing disaster to whole countries. It is why Western countries now struggle under the awful regimes imposed upon them by sexual extremists, and the ills they bring, killing millions. It is why we see a vast increase in murders and terrorism.

Where men do not care what happens to others, they will tolerate, or indulge in, death and killing. This is why environmentalism wants to see the deaths of millions by starvation, by abortion, or by 'natural' disasters.

> "Those who assault Western civilization in favour of ethnicity, primitivism, environmentalism and subjectivism, should realise their critiques and the alternatives they present are almost identical to those made by theorists of fascism. The fascists of the 1930's also sought to dismantle Western Civilisation and human-centred values. They too attacked the concept of individual identity and taught that reality is socially constructed; they too insisted that underlying all institutions is naked power; they too prized ethnicity; they too were environmentalists, They, too, questioned the objectivity of meaning." (134)

What we are seeing in Al Gore and his fast-moving demands is fascism mixed with Marxism. He is railroading individuals into becoming part of an unified whole (environmentalism), and railroading governments into doing the same. But, this suits governments, who are using the myths to rationalise their big-tax hikes and control of choices. They are even willing to destroy lives through support of delusional 'transgender' demands and homosexuality in general. The main task, though, is to firstly destroy Christians and biblical morality and ethics.

Though the science Gore presents has no basis in actual scientific fact, it is pushed anyway, because once a leader has unified his troops, they will do his bidding, regardless of what he is saying.

Al Gore is removing human-centred values and putting Gaia in their place; the worship of the inanimate. You do not need to start playing eastern music or call yourself a Gaiast; you do not need to attend a Mother Earth meeting or burn scented candles. Just follow Al Gore and environmentalists and they will do all that for you. Your genuine involvement is not required. All they want is your compliance, whether or not you agree with them or take part.

Panic Costs Dearly
By spreading panic in your own home and turning down your thermostat over mythical Green issues, you help the cause! Turn down your thermostat because your house is too warm, or because you want to save money. But, do not do it to comply with Green demands! In the UK there is a big push for citizens to install meters to check their use of energy. Even on TV ads the claim is that this saves the individual money by telling him his use of gas or electricity could be cut down. It is all part of Greenism... the common sense thing to do is simply to click off the light not being used, and to shut off the central heating if it is not being used! But, 'smart meters' are ways to spy on energy use and to cut down the bills of energy companies,

rather than to help the individual. It is a subtle Greenism most are unaware of.

The UK government is using massive and continuous propaganda techniques to make UK citizens compliant. Teenage Green 'activists' whose experience of the world is limited mainly to what they see on TV or on their games consoles, stand on podiums demanding compliance to Green objectives! They haven't got a clue about what they are demanding, but it is a 'cause', and teenagers love causes.

People are actually turning down their thermostats and buying those light bulbs with terrible light. They are buying cars with less CO_2 emissions, not because they wish to, but because it helps 'save the planet', or because government is making anything else very difficult. Governments now want us to comply by buying electric cars, even though the electrical supply infrastructure is not there!

When you buy new double-glazed windows one of the first marketing ploys used by the ignorant salesman is to tell you that your CO_2 footprint can be reduced! None of it is real! They do not have a clue! All I want to know is the price, and if the noise made by kids outside can be heard! Letting carbon out the window is irrelevant.

It is all one big scam. Environmentalism is only a superficial means to get people to comply with a

Marxist-fascist plot to destroy your culture and way of thinking! The fascism our fathers so valiantly fought has been re-introduced in the guise of Green ideology. The rest will follow.

If you have truly seen what Marxism did in the last century, then pull back! If you saw what happened during Hitler's day, then stop supporting Greens! You can be environmentally friendly without their help and pushing.

And because there is no actual need to panic, you can start to think for yourself again. Would not that be refreshing? Surely you want to do things better because of genuine science, rather than because someone is rushing you towards a precipice?

The 'New Agers' are now coming into their own. At one time the eccentrics hanging onto the fringes, they are now mainstream; any and all bizarre ideas are welcome. It is frightening for anyone who thinks independently to see objective science being replaced by pseudo-science based on Mother Earth worship!

Fascism Brings Deterioration

"Only a few prefer liberty – the majority seek nothing more than kind masters."(135)

"It is a terrible thing when you think you
got on a bandwagon and it turns out to
be a garbage truck." (135)

In 1830 Thomas Macauley said that whilst it may be
hard to prove people are in error when they tell us
our society has come to the end of its best era,
those before us said similar things, without tangible
reason to believe them.

He then asked, "On what principle is it that, when
we see nothing but improvement behind us, we are
to expect nothing but deterioration before us?"
Good point.

The Christianised past offends the pagan present,
so it must be dismantled and replaced by Gaia,
Marxism and fascism, despite their violent and anti-
human history. It does not make sense.

As has already been said, you can look at any
communist regime you wish, and you will see
deterioration, foul beliefs and practices, disregard
for humanity and common-sense, and rejection of
true science, based on a hatred for transcendence.

You can also look at Hitler's epitome of fascism. If
you like what you see, then you are a very sad
person, who must want fascism because it gives
you a purpose you do not presently have, one that
wants a world of vile disaffection and awful acts.

A man who asked to remain nameless recalls the isolationism of pro-Green scientists, as experienced through an old school pal:

"I (told him of my) decade of mucking through the theory, practice and historical origins of environmentalism. I've concluded that contemporary environmentalists are a band of frauds, and Third-Way-fascists, bent on reversing the Industrial Revolution and wrecking capitalism, and that Al Gore was one of them." (137)

Now, the old school buddy he was writing to evidently assumed this man (who we will call Mr X) had a PhD and wrote back accordingly, asking him if he would like to cooperate as a researcher.

Once the man replied that he did not have a PhD the old school buddy promptly stopped writing! This is an incredible but common sign of arrogant fascism! Having looked at some of the man's research I can say it ranks well against anything produced by PhD's.

The point I am making is that beforehand his research was fine, but when it was discovered he did not belong to the PhD clan, his work, though good, was rejected!!

I say this because you do not need to be a PhD to oppose environmentalism. You just need a good eye for the ridiculous and possibly an understanding of the research process. Even without any of this you have the right to ask questions or to express doubts or to reject intellectualised waffle.

Fascist Academics

Fascist academics, however, do not respect the minds of people who do not belong to the unified classes. They cannot be tolerated. That is what Hitler said and it is good enough for them. And we now find a little-Hitler around every academic corner, controlling minds and pass-rates of students.

The irate scientist argued that Mr X had "brutalised logic" using "dubious sources" that "slandered" Mr Gore. That is the kind of intolerance and garbled attack used by people who are following a guru. It is not the language of reason or scientific truth.

It often amuses me that people who think they know more than you tend to be very paternalistic or, more often, downright abusive. It is, as we have noted elsewhere, a sign of pseudo-science and the Marxist-fascist technique of telling and commanding people, rather than arguing your case.

It is also a sign that those who refuse to allow you to speak are very unsure of their ground, so they hit out.

As Mr X pointed out to his mild-mannered scientist buddy, "Few discoveries are more irritating than those which expose the pedigree of one's ideas" (139) quoting John Acton.

Get Funding by Lying
In the current scientific climate (good pun, eh?), in which scientists struggle to obtain funding, they resort to pseudo-scientific projects. Though said in wryness, it is true that all they have to do is insert the words 'global warming' or 'climate change' or 'CO2 emissions' into their funding applications and the funding will flow!

That is why they shout so loudly when anyone opposes them. It is why they claim a 'consensus' about the issues where none exists. It is a way of silencing everyone who knows the truth. And the public just believe it! Obviously, if the very foundation of one's work is removed or shown to be false, it brings down the entire edifice. Show that the claims made by Gore et al are false, and there is no longer a worldwide environmental industry, and it will come crashing down!

Hence, Gore demands that the world acts fast – he is terrified everyone will know him to be a liar and a fraud before he makes his huge profits. After he makes his profits? He will not care!

> "There are few signs yet that we have
> intellectual courage to admit to ourselves
> that we may have been wrong. Few are
> ready to recognise that the rise of
> fascism and Nazism was not a reaction
> against the social trends of the preceding
> period, but a necessary outcome of those
> tendencies."(139)

We have seen what this meant in the pre-Nazi regime of Victorian and Edwardian Germany, where everything in Germany was working towards something else – which turned out to be National Socialism. Seeds have a habit of growing, and what grows is characteristic of the seed.

The horrors of its existence were merely the outcome of the way of thinking of earlier men. The same author continues:

> "This is a truth which most people were
> unwilling to see even when the
> similarities of many of the repellent
> features of the internal regimes in
> communist Russia and National Socialist
> Germany were widely recognised. As a
> result, many who think themselves
> infinitely superior to the aberrations of
> Nazism, and sincerely hate all its

manifestations, work at the same time for ideals whose realisation would lead straight to the abhorred tyranny."(140)

I have already identified this peculiar blinkered view of life in today's Western societies. No doubt everyone (but neo-Nazis) would deplore fascism.

Yet, those who are willingly being dragged by the nose into Green policies and ideology are working towards a fascist regime! The trouble is, they do not know it yet. But, when they wake up they will see it too late.

It does not matter if you accept the decrees of fascism as taught by Gore and others. They do not care if you accept what they say, only that you do what you are told. And many of you are doing just that by believing the lies that continually come from their mouths, and doing their dirty work, shouting-down 'deniers'.

They do it by making you think your ideas are your own, and you follow them because you have a desire to see change through protecting the environment (because you have been fed propaganda), but have no idea if the science they present is true; an act of mass intellectual suicide. On the other hand, it is a natural result of modern society in all its violent anti-truth manifestations.

Lying by Images

Most people believe what the authority says because the authority says so and for no other reason. They have handed-over their will to liars and frauds, who will soon devour them.

> "The whole apparatus for spreading knowledge – the schools and the press, radio and motion pictures – (are) used exclusively to spread views which, whether true or false, will strengthen the belief in the rightness of decisions taken by the authority; and information that might cause doubt or hesitation be withheld." (140)

Earlier we saw that images are replacing words and proper arguments. Al Gore's film, 'An Inconvenient Truth', is filled with emotional shots of animals and ice crashing into the sea, cute penguins, etc. The aim is to use these images and emotion-inducing music, to convince viewers that what they see is genuine.

For this reason – and against all reason – viewers pass on his lies by saying "See, this is proof that global warming is happening right now, and that human beings are responsible for it."

What they should do is watch the film and remember it was a movie, not reality. It is just as rational as saying that Klingons are real because you saw them in Startrek!

This is too big a scenario to just accept Gore-isms without question! It is not proof of anything! One video clip does not make a whole, or even a part, of a scientific argument! Some of these images are discussed in the 2009 book. For now, just remember: a few seconds of film does not 'prove' anything. All you see is ice falling into the sea. This fall can be caused by any number of things, but you blindly accept it is 'caused' by man. Indeed, the SAME images passed off as 'genuine' by Gore over ten years ago are currently being recycled on TV and in the other media!

The idea that man is causing damage, is a matter of very complex science, and Al Gore's film opposes every proper method used by genuine science! Indeed, most of his film has been declared to be fraudulent, by a British court that cross-examined scientists from pro- and anti- warming camps.

The images portrayed have nothing to do with science. Any good film-maker can make you believe anything. That is a fact. That is why marketing works on TV. And the same marketing techniques are being used to make you believe there is a 'problem' that is a 'crisis'.

So, why are you, a non-scientist, (or a non-climate academic) blindly accepting what is said, and entering into panic mode? It does not make sense, and it questions your ability to come to logical conclusions. I do not say this easily or with the aim of making you feel stupid, because, by ignoring proper science and genuine facts, you are proving to be unreasonable and unthinking.

Al Gore has induced you to panic! He is a showman, not a scientist! Stop it right now and look again at the real science. Above all, realise that what you are being sucked into is fascism and Marxism, via a snake-oil salesman.

Al Gore and governments are deliberately foisting lies on the public in order to gain power and money. (This is proved in the 2009 book). They are holding back, and silencing, counter-arguments, using the media and children as channels of propaganda. They want you to believe every word they utter, with no questions asked. And it is why you must "act fast" to "save the planet". No time to ask questions or to query the research!!

Since when, in the past century, have people truly believed what governments and ad men say? So, why believe them now, without question, when the media are filled with the evils and lies told by politicians, often without any redress whatever? By complying without question you are aiding Al Gore and world governments to build a new regime. In return all you get is poorer and suppressed.

They have openly said they want to do this. And that their regime is a mix of Marxism and fascism, neither of which is acceptable or good for any race, creed or person.

Are You Borg, or Free?

If you are familiar with the Star Trek series, you will remember the Borg! They travelled about space in what resembled massive black cubes. Their aim was to subdue and assimilate all species of aliens, including human, into one mind, connected to everyone else by thought and electronics.

No Borg was an individual or was allowed to think separately. Everything was centralised and every thought was universally applied. Is this not Marxism?

Through this regime every captured person was subdued and absorbed, assimilated by the Borg, and whole civilisations were destroyed. Are you Borg? If you say "No"... Then why are you accepting Gore's version of events and reality, when he is manipulating you?

"Reality exists in the human mind, and nowhere else. Not in the individual mind, which can make mistakes, and in any case soon perishes; only in the mind of the Party, which is collective and immortal." (141)

The Borg must have got them! I readily admit I can make mistakes. But, being a part of a huge collective mind would not stop mistakes; it would only make those mistakes even bigger!

Think, my friends, before you are absorbed and cannot find your way out. Check all claims! Do not be Borg!

Schools Are Frontline
Al Gore has called for all California schools to teach 'the facts' about climate change. Even though the climate change Gore espouses does not exist. In the UK a survey taken soon after Gore began his antics found that half of all schoolchildren given his 'facts' could sleep at night because they were petrified they would die in a year or two. (142) I sincerely hope they saw sense when they grew up!

This survey result was because the UK government told all schools to show the Al Gore film, even though it's credibility has been seriously damaged by the UK courts and by other scientific facts! Government is growing its own supporters! Get them young and you can tell them to do anything. This is not about the environment, friends – it is outright political propaganda, put out by a Marxist government using fascist means. Sadly, even the current Tory government does the same thing.

Why are you allowing this to continue? Do you hate your kids? Why are you letting Al Gore and false science terrify your family into sleeplessness? How can this kind of mind terrorism be allowed? Easy. Government, Al Gore and environmentalism, are all fascist. They get what they want (power and extra taxation). Period.

"cumulatively, the American environmental movement is bread and butter for about 50,000 to 100,000 political workers." (143)

These work for a large number of environmental organisations, such as the infamous Sierra Club, National Audubon Society, World Wildlife Fund, Greenpeace, and so on. Each is an activist with families, friends and connections to many more sympathisers. So they have clout.

"anyone who believes there is no established party platform within the eco-movement would do well to survey the websites of the world's top 40 enviro-organisations. What one will uncover is a massive, integrated propaganda network distributing freight train loads of monotonously uniform Neo-Malthusian, Neo-Pagan, Neo-Luddite, and apocalyptic balderdash. There is a party line and a party press." (144)

"The movement also has a broader degree of organisational centralisation than most people realise. Whereas the seemingly disparate local eco-activist groups number in the thousands, these groups are financially dependant on, and are in the control of, about 100 funding agencies. These agencies consist principally of about three dozen philanthropic initiatives supplemented by an ever larger, even more activist EPA. (The cavalry of Young Turks in the Secretary of the Interior portfolio also merit attention).

Basically, when Carol Browner, Bruce Babbit, Ted Turner, Steven C. Rockefeller, and William Clay Ford Jr. sit down to a power lunch, the executive of the 'environmental movement/party' has convened." (144)

The article above was written just before the 2000 US elections. The commentator says "This year the 'Environmental Movement/Party' candidate for President is Al Gore." Remember, we have seen that this movement is Marxist and fascist to the core. So is Gore.

Throughout his political career Gore has been known for wearing an environmental/waste hat, so his current activities are no surprise. What makes his activities dubious is the way he is now so frontally fascist in technique... all supported enthusiastically by leading environmentalists (socialists).

If he continues to be treated as genuine, he will ruin America, as well as other countries. He is joined by the evergreen but dangerous Tony Blair, and the unscrupulous Clintons and Obama. All Marxists.

Apart from the obvious supporters, Gore has many more hidden away in the background, whose work gives him a blank cheque; people like the Society for Environmental Journalists and the North American Association of Environmental Educators. They all push the same regressive, oppressive ideas.

Then there are TV and radio presenters who belong to the Environmental Media Association. There are also thousands of rabbis, priests and (existential) ministers who write and disseminate eco-theology literature, endorsed by their religious hierarchies through the National Religious Partnership on the Environment. And some of these even produced a 'Green Bible'! (See my book critical of this deception. Available from Amazon and Lulu).

Between them they reach and influence over 100 million people! That is how Al Gore had such phenomenal success so rapidly and has recently resurrected his pack of lies and unscientific nonsense. He already has a massive following predisposed to his crank messages. So, he only needs to brainwash a relative few... the others have already brainwashed themselves.

Like any good fascist, Gore is causing all his followers to think they want what he wants. It then becomes 'their' idea, and so they work all the harder to prove their loyalty. The story behind Al Gore's political pals is fascinating... read it.

Remember, the article I am commenting on was written in 2000. Yet, the conclusion by the author is prophetic (and pathetic):

> "So the fix is in. This election is all about an environmentalist seizure of power, but the general public has yet to be informed of this, let alone be treated to an actual debate about its merits.
>
> The facts are incontrovertible. Al Gore is an eco-zealot. The environmental movement is united in praise around (him). The environmental movement is the main driving camp... Over the past 35 years the environmental movement has become so large, institutionalised and centralised, that it must be considered, for all intents and purposes, a third party in this election... Uneasy lies ahead." (144).

The writer of the article, William W Kay, was a lawyer, and he says Al Gore won his Vice-Presidency by demanding several recounts – unheard of in political history – because he did not like that he had lost in Florida to George Bush. The lawyer said that Gore pushed through a number of law-suits that were known as "frivolous and vexatious litigation' to get his own way.

It is a method used today by environmentalists (especially the Sierra Club) many times over, to get people to comply with their demands. Rather than enter into a court wrangle and huge costs, people/organisations just give in. Fascism!

And so Gore's pals get their way, time and again. In other words, when they get their way (and huge court pay-outs, considered to be easy money by extremist groups), Gore manages to silence even more members of the unbelieving public.

Kay says this kind of litigation is used to cripple critics financially, and that "it is not generally deployed as a means of seizing power." In the USA it is now used all the time to silence opponents of global warming. Kay says Gore's "green politics are extreme". And the seized cash is sued to spread more propaganda.

The eco-movements gather in over $10 billion annually!! The opportunity they have, and use, to brainwash the public, is staggeringly massive. And the longer a Democrat administration (USA) stays in power the worse it will get. Their aim is very simple – to destroy liberty and impose Marxism-fascism on the USA. And, as seen throughout 2016 to the present (2018), they will resort to violence to get it done. Indeed, it is no longer a resort but a plain wicked policy!

Continually, Democrat leaders write laws in secret and rush them through. The same happens in the UK. Fascists have taken over the constitution and smothered it in a blanket soaked in Marxism.

Student Savvy
The LaRouche Youth Movement says "Young Texans aren't buying Gore's Greenie Fascism" and want to "rid Texas of the putrid stench arising from Al Gore's recent documentary: 'An Inconvenient Hoax'." This is mentioned to show there are still some young people in America who can think for themselves, and know Gore's greenism is fascist scientism.

The LaRouche group met with University of Houston students, to tell them about the "true cause of climate change as well as the nefarious motives" of those backing Gore's "most recent propaganda effort for planetary depopulation." The resultant meeting proved to be quite rattled, with the usual irrationality found amongst Gore followers, the older 'Baby Boomers'.

The meeting even included, to my surprise, mention of the "change in culture within the US and Europe that produced our current crop of morality and scientifically-challenged Baby Boomers... who are impulsively jumping on the Gore bandwagon." (145) The speaker "pinpointed the Anglo-Dutch oligarchy, as typified by Tony Blair... as currently using Gore's intentional fraud to help accomplish their objective of drastic population reduction through the shutdown of what little real industrial capability remains... youth must have a commitment to truth."

The assessment by LaRouche was that Baby Boomers, the older students and staff, were behind Al Gore, because that is what sheep do best. But, younger students "are far less committed to supporting a policy whose unavoidable outcome is their own annihilation", as they recognise the fascist doctrine "being peddled by the likes of Gore and his Anglo-Dutch backers." (145)

Referring to a New York Times editorial 'Moving Beyond Kyoto', and a corresponding co-ed article, 'Alas, Al Gore Got it Wrong', Marcus Epstein talks of 'Al Gore's Global Green Fascism'. (146) Many writers and articles all come to the same conclusion, that Al Gore is fascist and so are his ideas. This is because he is fascist! But, he is also Marxist. The combination is frightening.

Remember what Hillary Clinton said on her irate and nasty campaign to get elected at any cost:

"We're going to take things away from you on behalf of the common good." (147). And yet many voted for her!

Hillary (like Obama) is a great Marxist with a lovely nasty streak. Her idea is simple – tax richer folks until it hurts and give it to the poor... including millions of illegal immigrants she refuses to kick out of the country! Now that is solid Marxism.

We have taken a brief look at how Marxism and fascism are at the root of the modern environmental movement. In this chapter we see these movements have been brought right up to date, under Al Gore and the lies he continually feeds to the gullible and people who have now, sadly, found a purpose in their lives, despite the fact that the purpose is built upon deception, fraud, and socio-spiritual harm.

If you read the 2009 book, you will find Marxism and fascism coming up time and again, not because it is being used to slander or call names, but because Marxism and fascism are the foundation of modern environmentalism. Ignore the facts at your peril, because having a romantic notion of communism and Nazi thought in a relatively 'free' country in the West, is very different from actually living out these vile movements in reality. Friends, you are being duped; reclaim your minds and stop supporting Gore!

Note: Prof George Sessions referred to 'Earth Day', 1990, as a "minor upward blip in what has been a steady decline of the ecology movement since 1981." (148) In reality, it declined because its ideas were nonsense. On the other hand, slick propaganda has turned this event into a big-time attraction.

Even the hotel industry got in on the act, voraciously and irresponsibly spreading Goresque deception in its publications. (149) Though I published a rejoinder (150), it was like dropping a spoonful of water into the ocean! Hotels and resorts just do not want to know – they can only see the profits from cashing-in on the global warming scare and pretending to be 'Green'. It is all about cash-flow, Not the 'earth'.

At that time corporations tore down the day's arguments. Of course, Greens saw this as a sign of capitalist thuggery. In reality, it was businesses looking after their interests and those of national well-being. Trump has also shown a hard-headed view of Green lies, and the world should thank him for his lead.

Then, after Earth Day 25, 1995, Sessions saw this as "an even bigger flop" for the same reasons. But, there is no point in saying this was caused by capitalist propaganda when Greens use nothing but propaganda all the time!

"Earth Day 25 essentially mirrored the confusion and disarray of the contemporary environmental/ecology movement."

Sessions referred to a Republican backlash against Green ideas that would have taken the USA back in time and down into financial instability. He also said that, at that time, membership of extremist groups such as the Sierra Club went down by 20% - 30%. Scaring people only lasts for so long! When they see the sky really isn't falling down, they relax and see the Green movement for what it is – fraudulent.

This is an interesting backwards-look at the situation just over 10 years ago, because, at exactly the same time (but nothing to do with ecology), an awful cultic religious movement swept the world, called the Toronto Blessing. It literally destroyed thousands upon thousands of churches by turning members towards a fake miracle movement.

The movement itself is of no consequence to this book. What is of interest is why the Toronto Blessing arose in the first place. Like the Green movement, charismaticism was on the wane.

So, charismatic church leaders decided another fantasy was needed to give them a big boost. They came up with amazing but fake miracles, and the whole thing took off, under leaders who could make stone come alive through occultism!

This is what has happened to the Green movement. It was naturally dying because it had nothing to offer the world, so needed a big boost. Along came Al Gore! Just as the charismatic church was only a fringe 'nutcase' movement until the 'new wave' was invented, so Gore was sidelined as Vice-President and lost his office. Now, he is to be resurrected, embalming fluid washed off, to lead a new Green assault.

Do not under or over-estimate the Green movement. It will do everything in its power to drive dissidents into the ground. Just ten short years ago businesses fought the Greens. Now, they embrace them. It took Al Gore's film to make Green ideas seem truthful. No truth, just celluloid. It's all fake, an era of fashionable accessories, a pathetic 'cause'; to have because people have little else in their lives.

Goodbye reality - Hail Hollywood!

Late Note: On the day I uploaded this book to the publisher, the WWF televised a reminder of all the above, by calling viewers to save the planet! The language used was different from the language they used ten years ago, but the aim is the same – to make people think the planet is in grave danger, so they must choose to save it!! What audacious and political garbage! As usual they used great photography… but their argument was fallacious!
Don't be duped!

References

Chapter one
1.   'Truth is My Sword', Volume 1, as quoted in tparents.org
2.   BBC address, 26th March, 1979
3.   Wikipedia.org
4.   21st August, Friedrich Braun, Economics and Finance, quoted in thecivicplatform.com
5.   'The Green Movement: A Socio-Historical Explanation', Galtung. Sage Publications, London, P235, in 'Globalisation, knowledge and Society', Ed. Albrow and King. Originally presented as a lecture at FLACSO, Santiago, Chile, December, 1984
6.   P236, the Green Movement
7.   'Comparative Green Politics: Beyond the European Context?' PR Hay and MGHaward, Political Studies, Vol. 36, Sept. 1988, P433-448. Blackwell Synergy
8.   'Green Confucianism: Ecology, Class and the Green Movement', Patrick Eytchison, Redwood Coast Greens, in Synthesis/Regeneration 31, Spring 2003, quoted on greens.org
9.   'Green Religion as Science', 23rd August, 2001, Doug Casey, worldnetdaily.com
10. Source: 'Environmentalism and Political Theory. Towards an Ecocentric Approach', Robyn Eckersley, publ. State University of New York Press, 1992
11   Eckersley, Page 4
12. 'Marx', 1981: p958-9

13. 'The International Handbook of Environmental Sociology', Chapter 12: 'Beyond Sociology: Marxism and the Environment', Peter Dickens. Edward Elgar Publishing, 1997

14. 'Marx's Theory of Metabolic Rift: Classical Foundations for Environmental Sociology', John Bellamy Foster, university of Oregon, American Journal of Sociology, Vol 105, No 2, 1999. This book is recommended by socialists

15. 'Marxism: An Historical and Critical Study', George Lichtheim, New York, 1964

16. For criticisms see 'Marx's Theory of Metabolic Rift', J. B. Foster, American Journal of Sociology, Vol 105, No 2, 1999, pp366-405

17. Capitalism, Nature, Socialism Journal

18. 'Marx's Ecology in Historical Perspective', J B Foster, International Socialism Journal, Winter 2002

19. Human Events

20. Source: 'Capital', K Marx, Vol 1, New York, 1976, p 638

21. 'Capital'. Vol 3, 1981, pp 948-950,959

22. Appendix 7

23. 'Marx's ecology in historical perspective', J B Foster, International Socialism Journal, Winter 2002, Issue 96

24. Capital, K Marx, Vol 3, New York, 1981, pp 948-950

25. Capital

26. Capital, p 911

27. Capital, Vol 1, p637; Vol 3, p959

28. socialistreviewindex.org.uk:80

29. socialistreviewindex.org.uk:80

30. socialistreviewindex
31. The German Ideology', K Marx
Chapter 2
32. Cape and Islands News, March 13th, 2008
33. Forest Clayton, 'Suppressed History', Armistead Publishing, Ohio, 2003, p 8. Clayton has written three books in a series; well worth reading
34. thecivicplatform.com
35. Review of 'How Green Were the Nazis?' Nature, Environment and Nation in the Third Reich'. Brüggemeier, Mark Cioc, Thomas Zeller, Publ. Ohio University Press
36. 6th May, Freidrich Braun, thecivicplatform.com
37. Dan Gainor, Business and Media Institute
38.'Blood and Soil: Richard Walther Darre and Hitler's "Green Party", Anna Bramwell, Bourne End: Kensal, 1985 and her 'Ecology in the 20th Century: A History', New Haven: Yale University Press, 1989 et al
39. h-net.org/reviews
40. h-net, p 4
41. Review by Wilko Graf von Hardenberg, of 'How Green Were the Nazis? Nature, Environment, and Nation in the Third Reich', Brueggemeler, Cioc, Zeller, eds. Athens: Ohio University Press, 2005
42. Hardenberg, P14
43. 'Demythologising Environmentalism', Douglas Weiner, Journal of the History of Biology 25, No. 3, 1992. Pp385-411.
44. Weiner, P245
45. leooshkosh.blogspot.com

46. 'Hitler's Table Talk', Hitler, 11th November, 1941, Section 66. As in book by same name, Martin Bormann. Available from Amazon

47. Heinrich Himmler quoted in Heinz Haushofer, 'Ldeengeschichte der Agrarwirtschaft und Agrarpolitik im deutschen Sprachgebiet, Band II, München, 1958, p 266

48. G Gordon Liddy: Volce of Unreason, November 22nd, 2004

49. Ernst Lehmann. 'Biologischer Wille. Weg und Ziele biologischer Arbeit im neuen Reich. München, 1934, p 10-11

50. 'The Environmental Movement in Germany: Prophets and Pioneers, 1871-1971', Raymond H Dominick, Bloomington, 1992

51. Quoted in Rudolph Krügel, 'Der Begriff des Volksgeistes' in Ernst Moritz Arndt's 'Geschichtsanschauung', Langensalza, 1914, p 18

52. spunk.org

53. Wilhelm Heinrich Riehl, 'Feld und Wald', Stuttgart, 1957, p 52

54. Klaus Bergmann, 'Agrarromantik und Großstadtfeindschaft', Meisenheim, 1970, p 38. Großstadtfeindschaft is a word meaning hostility to 'consmopolitanism, internationalism and intolerance of cities'

55. spunk.org

56. George Mosse, 'The Crisis of German Ideology: Intellectual Origins of the Third Reich', New York, 1964, p 29

57. Lucy Dawidowicz, 'The War Against the Jews, 1933-1945', New York, 1975, pp 61-62

58. spunk.org
59. Daniel Gasman, 'The Scientific Origins of National Socialism: Social Darwinism in Ernst Haeckel and the German Monist League', New York, 1971, p 17
60. Gasman, p 30
61. spunk.org
62. Gasman, p 33
63. Foreword, 1982 reprint: Raoul Francé, 'Die Entdeckung der Heimat', MUT Verlag, 1923
64. spunk.org
65. spunk.org
66. spunk.org
67. Ulrich Linse, 'Okopax und Anarchie. Eine Geschichte der okologischen Bewegungen in Deutschland, Munchen, 1986, p 60
68. Peter Staudenmaier, 'Fascist Ecology: The 'Green Wing' of the Nazi Party and its Historical Antecedents.', spunk.org
69. spunk.org
70. spunk.org
71. spunk.org
72. spunk.org
73. Michael Zimmerman, 'Heidegger's Confrontation with Modernity: Technology, Politics and Art', Indianapolis, 1990, pp 242-3
74. spunk.org
75. spunk.org
76. spunk.org
77. Robert Pois, 'National Socialism and the Religion of Nature', London, 1985, p 40, and p42-3
78. Associated Press, 24th September, 2006

79. Quoted in 'Toxic Terror'
80. Quoted in 'Toxic Terror'
81. (Paul Watson, quoted in Access to Energy, Vol 10. No 4, December 1982).
82. Henry Picker, 'Hitlers Tischgespräche im Führerhauptquartier 1941-1942', Stuttgart, 1963, p 151
83. Adolph Hitler, 'Mein Kampf', München, 1935, p 314
84. Gert Gröning and Joachim Wolschke-Bulmahn, 'Politics. Planning, and the Protection of Nature: political abuse of early ecological ideas in Germany, 1935-1945', Planning Perspectives 2, 1987, p 129
85. spunk.org
86. Änne Bäumer, 'NS-Biologie, Stuttgart, 1990, p 198
87. spunk.org
88. spunk.org
89. Picker, 'Hitlers Tischgespräche', pp 139-140
90. spunk.org
91. R Walther Darré, 'Um Blut und Boden: Reden und Aufsätze', München, 1939, p 28
92. Bramwell, 'Ecology in the 20th Century', p 203
93. spunk.org.
94. Anna Bramwell, 'Darré. Was This Man 'Father of the Greens'?', History Today, September 1984, Vol 34, pp7-13
95. spunk.org
96. A term used by Fritz Todt, "one of the most influential National Socialists", Franz Neumann,

'Behemoth. The Structure and Practice of National Socialism', 1933-1944', New York, 1944, p 378

97. Alwin Seifert, 'Im Zeitalter des Lebendingen', Dresden, 1941, p 13

98. Raymond Dominick, 'The Nazis and Nature Conservationists', The Historian, Vol XLIX, No 4, August 1987, p 536

99. Dominick, p 536

100. spunk.org

101. Gröning and Wolschk-Bulman, 'Politics, Planning and the Protection of Nature', p 137

102 Gröning, ibid p 138

(Note: I am grateful to 'Ecofascism: Lessons from the German Experience", and Janet Biehl & Peter Staudenmaier, for their article, 'Fascist Ecology: The Green Wing of the Nazi Party and its Historical Antecedents', which I have used as a source. To read the full extensive article, go to spunk.org)

102a Bramwell, 'Ecology in the 20th Century', p 48

Chapter 3

103. Interview with Jonah Goldberg, 'Liberal Fascism'

104. 8th September, 2005, tubearoo.com

105. John Bellamy Foster, 'Marx's Ecology: Materialism and Nature', Monthly Review Press, 2000

106. monthlyreview,org

107. monthlyreview.org

108. monthlyreview.org

109. 7th December 2007, 'Marx and the Global Environmental Rift', greenleft.org.au

110. Brett Clark and Richard York, 'Carbon Metabolism: Global Capitalism, Climate Change, and the Biospheric Rift'. Theory & Society, Vol 34, 2005, p 419

111. Foster

112. Foster

113. Martin W Lewis, 'The Capitalist Imperative' in 'Green Delusions: An Environmentalist Critique of Radical', Duke University Press, 1994

114. Bookchin 1989:128

115. J O'Connor, 1989b:10) (p 151 Green Delusions

116. O'Connor, p 152

117. Jonah Goldberg, 'Liberalism and Fascism: The Secret History of the American Left, from Mussolini to the Politics of Meaning', Doubleday, 2008

118. Book World, Washington Post

119. January 8th, Reed Business Information

120. 'Why Green is Red: Marxism and the Threat to the Environment', isjtext.ble.org.uk

121. Joel Kovel, MD, Alger Hiss Professor of Social Studies, Bard College: Review of 'The Greening of Marxism', Ed Ted Benton, The Guildford Press, 1996

122. Interdisciplinary Studies in Literature and Environment

123. Above three quotes from 'Green Politics in Impasse', Lisa Macdonald, dsp.org.au

124. 'Political Ecology and the Future of Marxism', lipietz.net

125. Political Ecology

126. 'Modern Fascism', Gene Edward Veith jr., p 25

127. Modern Fascism, p 16
128. 'Fascism', Bill Crouse, CIM Outline 54, rapidresponsereport.com
129. Fascism, p 2
130. Fascism, p 4
131. Veith, p 12
132. Veith, pp 12,13
133. Veith, p 13
134. Veith, p 22
135. Sallust. 'Histories'
136. Ernst 'Putzi' Hanfstaengl
137. lewrockwell.com
138. lewrockwell, p 2
139. Hayek, 'Road to Serfdom'
140. Havek
141. G Orwell, '1984'
142. 'Liberal Global Warming Fascists Brainwash Elementary Students', YouTube
143. William W Kay, 'The Gore Presidential Bid', Environmentalism is Fascism, ecofascism.com
144. The Gore Presidential Bid
145. 'Young Texans Aren't Buying Gore's Greenie Fascism', Lyndon LaRouche Political Action Committee, larouchepac.com
146. 3rd July, 2007, infowards.net
147. 22nd February 2007, The Washington Times
148. George Sessions, The Trumpeter, Vol 12, No 3, 1995
149. Hotel Interactive, 22nd April, 2008 and Hotel-Online, 2nd April, 2008
150. 4Hoteliers, 28th April, 2008

Also see related books by K B Napier:

'The Global Green Agenda, Second Edition'
(published 2009, but still highly relevant)

'Keep Off the Grass' (Publ. 2008: supplement to the
2009 book)

'Freezing – the New Warming' (Publ. 2009:
supplement to the 2009 book)

'The Green Bible – Bad Science. Bad Theology' (a
critical analysis: 2009).

*For theology books by K B Napier, go to
website below:*

Bible Theology Ministries
PO Box 415
Swansea
SA5 8QW
UK

Website:
www.christiandoctrine.com

Website email address for K B Napier:
napierkb@christiandoctrine.com

www.ingramcontent.com/pod-product-compliance
Lightning Source LLC
Chambersburg PA
CBHW070027210526
45170CB00012B/217